T0211342

BestMasters

Mit „BestMasters" zeichnet Springer die besten Masterarbeiten aus, die an renommierten Hochschulen in Deutschland, Österreich und der Schweiz entstanden sind. Die mit Höchstnote ausgezeichneten Arbeiten wurden durch Gutachter zur Veröffentlichung empfohlen und behandeln aktuelle Themen aus unterschiedlichen Fachgebieten der Naturwissenschaften, Psychologie, Technik und Wirtschaftswissenschaften. Die Reihe wendet sich an Praktiker und Wissenschaftler gleichermaßen und soll insbesondere auch Nachwuchswissenschaftlern Orientierung geben.

Springer awards "BestMasters" to the best master's theses which have been completed at renowned Universities in Germany, Austria, and Switzerland. The studies received highest marks and were recommended for publication by supervisors. They address current issues from various fields of research in natural sciences, psychology, technology, and economics. The series addresses practitioners as well as scientists and, in particular, offers guidance for early stage researchers.

More information about this series at https://link.springer.com/bookseries/13198

Simon Seelig

Characterizing Groundwater Flow Dynamics and Storage Capacity in an Active Rock Glacier

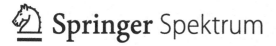 Springer Spektrum

Simon Seelig
Graz, Austria

ISSN 2625-3577   ISSN 2625-3615   (electronic)
BestMasters
ISBN 978-3-658-37072-5   ISBN 978-3-658-37073-2   (eBook)
https://doi.org/10.1007/978-3-658-37073-2

Responsible Editor: Marija Kojic
This Springer Spektrum imprint is published by the registered company Springer Fachmedien
Wiesbaden GmbH, part of Springer Nature.
The registered company address is: Abraham-Lincoln-Str. 46, 65189 Wiesbaden, Germany

The original version of the book was revised: Author name has been corrected. The correction to the book is available at https://doi.org/10.1007/978-3-658-37073-2_8

Acknowledgments

The completion of this thesis would not have been possible without the professional and personal assistance and guidance of many teachers, supporters and friends.

I would like to express my special thanks to my advisors Gerfried Winkler and Steffen Birk for many inspiring discussions and valuable suggestions, for their patience, guidance, and examination that enabled and shaped this thesis.

I am deeply indebted to Thomas Wagner and Magdalena Seelig for many hours of discussion and deliberation, for their important contributions to the concepts, results, and interpretations that make up this document and for their constructive critical review and corrections.

Particularly helpful to me were the propositions and comments by many friends and colleagues who took the time and applied a lot of effort to assist in solving problems and drawing conclusions. I wish to express my special thanks to Markus Eder, Raoul Alexander Collenteur, Matevž Vremec, Marcus Pauritsch, Andreas Pilz, Kajetan Heigert, Thomas Pedevilla and Rudolf Philippitsch.

This thesis builds on the data and results collected by several people during the last years. I would like to extend my sincere thanks to Till Groh and Jan Blöthe for providing the rock glacier surface velocity data. Many thanks to Michael Grimm, Anja Klebelsberg and the Tiroler Wasserkraft AG for providing meteorological data recorded at the nearby weather station Weißsee. The digital elevation model was kindly provided by the GIS Service of the Government of Tyrol.

Finally, I gratefully acknowledge the support and encouragement of my parents Harald and Frieda Kainz during my studies. I would also like to thank my friends who accompanied and heartened me during these years.

Abstract

Alpine aquifers play a critical role in the hydrology of mountainous areas by sustaining base flow in downstream rivers during dry periods and retarding flood propagation after heavy precipitation events. Progressing climate change alters climatic and meteorological boundary conditions as well as the hydraulic response of alpine catchments by ablating glaciers and thawing permafrost. Rock glaciers exert a controlling influence on the catchment response due to their prominent groundwater storage and complex drainage characteristics. This thesis investigates the hydrogeology and internal structure of the active rock glacier Innere Ölgrube (Ötztal Alps), which governs catchment runoff and is affected by permafrost degradation. A 3D geometrical model of its internal structure is obtained by combining geophysical data and permafrost creep modelling. Available data and new results are integrated into a conceptual hydrogeological model providing a sound basis for the implementation of a prospective numerical groundwater flow model. Hydraulic properties of the hydrostratigraphic units constituting the rock glacier are estimated and groundwater recharge fluxes quantified. Fundamental properties of the heterogeneous groundwater flow system within the rock glacier are discussed and compared to existing rock glacier studies.

Keywords: Active rock glacier · Groundwater flow · Storage capacity · Hydraulic properties · Conceptual model · Austria

Kurzfassung

Alpine Aquifere nehmen eine hydrologische Schlüsselrolle in Gebirgsregionen ein. Während Trockenzeiten gewährleisten sie den Basisabfluss der unterhalb gelegenen Bäche und Flüsse, nach Starkniederschlägen verzögern sie den Hochwasserabfluss. Der fortschreitende Klimawandel verändert sowohl die klimatischen und meteorologischen Randbedingungen als auch das hydraulische Verhalten alpiner Einzugsgebiete durch Abschmelzen von Gletschern und Abtauen von Permafrost. Aufgrund ihrer ausgeprägten Speicherfähigkeit und komplexen Grundwasserströmung im Inneren prägen Blockgletscher das Abflussverhalten alpiner Einzugsgebiete. Diese Arbeit untersucht die hydrogeologischen Eigenschaften und interne Struktur des aktiven Blockgletschers Innere Ölgrube in den Ötztaler Alpen. Der Blockgletscher dominiert den Abfluss des dahinter liegenden Einzugsgebiets. Sein gefrorener Kern ist von fortschreitendem Permafrostabtau betroffen. Die Kombination geophysikalischer Erkenntnisse mit einem mechanischen Modell ermöglicht die Konstruktion eines dreidimensionalen geometrischen Modells der internen Blockgletscherstruktur. Bereits verfügbare Daten werden um neue Ergebnisse ergänzt und zu einem konzeptionellen hydrogeologischen Modell zusammengeführt, das eine solide Basis für ein zukünftiges numerisches Grundwasserströmungsmodell bildet. Die Arbeit enthält eine Abschätzung der hydraulischen Eigenschaften der wichtigsten hydrostratigraphischen Einheiten, welche den Blockgletscher aufbauen, sowie eine Quantifizierung der Grundwasserneubildungsraten. Die wesentlichen Eigenschaften des heterogenen Strömungssystems werden diskutiert und mit bereits bestehenden Blockgletscherstudien verglichen.

Schlagwörter: Aktiver Blockgletscher · Grundwasserströmung · Speicherkapazität · Hydraulische Eigenschaften · Konzeptionelles Modell · Österreich

Contents

Symbols

a	dispersivity [m]
A	creep parameter [$\mathrm{Pa^{-n}\ s^{-1}}$]
b	saturated thickness [m]
B	discharge contribution at the beginning of the spring flow recession [$\mathrm{m^3\ s^{-1}}$]
c	tracer concentration [$\mathrm{mg\ m^{-3}}$]
CI	coefficient of variation [$-$]
c_p	peak concentration [$\mathrm{mg\ m^{-3}}$]
d	representative length [m]
D	dispersion coefficient [$\mathrm{m^2\ s^{-1}}$]
D_{Ch}	Chatwin estimate of dispersion coefficient [$\mathrm{m^2\ s^{-1}}$]
D_{MM}	method of moment estimate of dispersion coefficient [$\mathrm{m^2\ s^{-1}}$]
$\dot{e}$	strain rate [$\mathrm{s^{-1}}$]
$\dot{e}_e$	effective strain rate [$\mathrm{s^{-1}}$]
g	gravitational acceleration [$\mathrm{m\ s^{-2}}$]
h	thickness [m]
h_{AL}	active layer thickness [m]
h_{PF}	permafrost layer thickness [m]
k	permeability [$\mathrm{m^2}$]
K	hydraulic conductivity [$\mathrm{m\ s^{-1}}$]
ks	kurtosis [$-$]
l	distance [m]
m	match position [$-$]
n	creep exponent [$-$]
N	number of analyzed frequencies [$-$]
p_0	false alarm probability [$-$]

P_e	Peclet number [−]
P_N	normalized power [−]
q	specific discharge [m s^{-1}]
Q_0	discharge at the beginning of the spring flow recession [m^3 s^{-1}]
Q	discharge [m^3 s^{-1}]
R^2	coefficient of determination [−]
Re	Reynolds number [−]
r_m	correlation coefficient at match position m [−]
s^2	variance
sk	skewness coefficient [−]
t	time [s]
T	tortuosity [−]
$\langle t \rangle$	mean residence time [s]
t_p	time to peak [s]
t^*	test statistic [−]
$\dot{u}$	displacement rate [m s^{-1}]
$\dot{u}_{x,b}$	basal displacement rate [m s^{-1}]
$\dot{u}_{x,s}$	surface displacement rate [m s^{-1}]
v	linear velocity [m s^{-1}]
V	volume [m^3]
$\langle v \rangle$	mean linear velocity [m s^{-1}]
v_p	peak velocity [m s^{-1}]
V_m	water volume constituting mobile zone [m^3]
w	aquifer width [m]
$\langle y \rangle$	mean of time series Y
Y	time series
y_j	data point of times series Y at time t_j
α	recession coefficient [s^{-1}]
β	dimensionless partition coefficient for mobile and immobile zones [−]
γ	dimensionless response time [−]
δ	slope angle [−]
ζ	first-order mass transfer coefficient [s^{-1}]
η	dynamic viscosity [Pa s]
θ_m	mobile zone volumetric water content [−]
θ_{im}	immobile zone volumetric water content [−]
ϑ	water temperature [°C]
λ_a	characteristic aquifer response timescale [s]
λ_r	characteristic recharge timescale [s]
ρ	density [kg m^{-3}]

ρ_{AL} active layer bulk density [kg m^{-3}]

ρ_{PF} permafrost layer bulk density [kg m^{-3}]

ρ_w water density [kg m^{-3}]

σ stress [Pa]

ς_0 power level [$-$]

τ deviatoric stress [Pa]

τ_e effective deviatoric stress [Pa]

ϕ porosity [$-$]

φ hydraulic head [m]

ψ_1 proportionality constant (creep model) [Pa^{-1} s^{-1}]

ψ_2 proportionality constant (advection dispersion model) [mg s$^{0.5}$ m^{-3}]

Ω dimensionless mass transfer coefficient [$-$]

ω angular frequency [s^{-1}]

Introduction

1

Alpine catchment characterization is the key to evaluating the hydrological response of headwater catchments and predicting the hydraulic regime of downstream rivers and aquifers. The Alpine climate has been subject to significant changes during the last decades and is expected to be affected by the global climate response to increasing greenhouse gas concentrations during the 21st century (Gobiet et al., 2014). River peak runoff is predicted to shift to winter and early spring, since less precipitation will fall as snow and the snow cover will start to melt earlier (Barnett et al., 2005). The frequency and intensity of extreme precipitation events including heavy rainfall as well as drought conditions is predicted to increase (Rajczak et al., 2013; Gobiet et al., 2014; Ban et al., 2015). In crystalline catchments, rainfall is quickly transferred to the outlet by surface runoff. Sediment accumulations such as talus, moraines, or rock glaciers act as critical buffer by retarding flood propagation and sustaining spring flow during dry periods (Clow, 2003; Hood and Hayashi, 2015; Hayashi, 2020; Wagner et al., 2020b).

Active rock glaciers are slowly deforming ice-debris mixtures creeping downhill. Since their storage (buffer) capacity depends on the available pore space, their hydrologic response to precipitation events is controlled by the amount of ice filling the interstitial pores. Continuous permafrost forms an impervious surface inhibiting infiltration into deeper layers (Krainer and Mostler, 2002; Geiger et al., 2014). Progressing degradation of permafrost ice increases storage capacity and alters rock glacier and catchment runoff characteristics (Jones et al., 2019). In the Austrian Alps, an area of ~ 1280 km^2 is drained through rock glaciers, i. e. 13.7% of the total land surface area above 2000 m a. s. l. (fig. 1.1; Wagner et al., 2020a,c). Due to their ubiquitous presence, rock glaciers exert a controlling influence on the alpine groundwater system (Wagner et al., 2020a,b). Thus,

© The Author(s), under exclusive license to Springer Fachmedien Wiesbaden GmbH, part of Springer Nature 2022
S. Seelig, *Characterizing Groundwater Flow Dynamics and Storage Capacity in an Active Rock Glacier*, BestMasters,
https://doi.org/10.1007/978-3-658-37073-2_1

the ongoing degradation of permafrost throughout the Alps is likely to alter the hydrologic regime in mountainous regions (Haeberli and Beniston, 1998; Harris et al., 2003; Roer et al., 2008; Rogger et al., 2017; Beniston et al., 2018; IPCC, 2019). Predicting mountain catchment responses to climate change thus requires a process-based understanding of groundwater flow and storage components of rock glaciers.

As one of the best studied rock glaciers in the Austrian Alps, the active rock glacier Innere Ölgrube provides a well suited opportunity to study these processes. Its proximity to the lower permafrost boundary and two degrading cirque glaciers within the rock glacier catchment reflect a typical high alpine setting of active rock glaciers exhibiting large catchments. Several previous studies assembled a comprehensive data base characterizing its most important features. The aim of this thesis is to integrate all available data into a consistent conceptual and 3D geometrical model of its internal structure and hydraulic properties. The results are expected to provide a sound basis for the implementation of a numerical groundwater flow model.

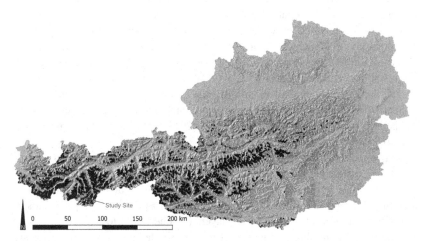

Figure 1.1 Rock glaciers in the Austrian Alps (Wagner et al., 2020a). The depicted rock glacier inventory covers the federal states of Carinthia, Salzburg, Styria, Tyrol, and Vorarlberg. The study site is located in the Ötztal Alps of Northern Tyrol (digital elevation model: Open Data Österreich, https://www.data.gv.at, Creative Commons Attribution 4.0 International (CC BY 4.0); rock glacier inventory: Wagner et al., 2020c, https://doi.pangaea.de/10.1594/PANGAEA.921629, CC BY 4.0).

Study Site

2

The active rock glacier Innere Ölgrube (OEG) is located in a small west facing side valley in the upper Kaunertal (Ötztal Alps, Austria) (fig. 2.1). Its alpine 1.83 km² catchment constitutes a headwater tributary to the Fagge River. It is characterized by rugged topography largely surrounded by steep cliffs except to the West, where the hanging valley abruptly declines to the Kaunertal. Its pronounced relief is reflected in altitude ranging from 2394 m to 3295 m a. s. l. (mean altitude 2887 m a. s. l.) and a mean slope angle of 36.5°. The terrain is composed of outcropping bedrock (38 %) mainly exposed on steep cliffs, while talus and debris slopes (23 %) make up most of the valley sides. The central and lower parts of the catchment are dominated by moraine (17 %) and rock glaciers (13 %). Two small cirque glaciers (9 %), both called Hinterer Ölgrubenferner, are present in its uppermost part, shadowed by north facing cliffs of the highest peak within the catchment (Hintere Ölgrubenspitze, 3295 m). Their meltwater reaches the rock glacier partly as surface runoff forming a small creek below staggered moraine walls before infiltrating in the rock glacier rooting zone. The remaining meltwater infiltrates into the coarse debris accumulations covering the valley bottom, reaching the rock glacier as subsurface flow. Bedrock geology is dominated by fractured polymetamorphic gneisses and schists forming a part of the Ötztal-Stubai Complex (Hoinkes and Thöni, 1993). Schistosity generally strikes E-W, steeply dipping to the south (Rieder, 2017).

Figure 2.1   Field impression of the rock glacier Innere Ölgrube and its associated catchment
(Photo: Gerfried Winkler).

The Ötztal Alps are part of the inner alpine climate province characterized
by dry conditions. The OEG catchment is characterized by strong temperature
gradients (due to its strong relief), a pronounced seasonal variation in meteorological conditions, and influence from both the Atlantic Ocean as well as the
Mediterranean Sea (Heigert, 2018). The catchment receives a mean annual precipitation of 800 mm throughout the study period (2013–2018). Mean monthly
air temperature is −10.6°C in January and 7.2°C in July, with daily averages
ranging from −24.5°C to 14.2°C. Snow cover lasts partly for 9 to 10 months per
year. Air temperature and precipitation intensity are recorded at the meteorological station Weißsee operated by Tiroler Wasserkraft AG. The records are taken
approximately 4 km from the rock glacier, at an elevation of 2464 m a. s. l.. Mean

catchment air temperature is estimated assuming a vertical temperature gradient of $5 \cdot 10^{-3}\,°C\ m^{-1}$, while precipitation intensity is assessed applying an elevation factor of $7 \cdot 10^{-4}\,°m^{-1}$, consistent with Wagner et al. (2020b). The low temperatures along with the predominance of bare rocks and coarse grained sediments within the catchment, sparse vegetation and a thin or absent soil cover imply low evapotranspiration rates.

The $0.24\ km^2$ OEG rock glacier is located at a mean altitude of 2582 m a. s. l., sloping westwards at a mean angle of 23.8°. Its tongue-shaped morphology, exhibiting a maximum length of 896 m at a mean width of 266 m, is segmented in two elongated lobes (fig. 2.2): The northern lobe consists of greyish weathering orthogneiss and exhibits a surface morphology characterized by pronounced longitudinal and transversal ridges and furrows. This part of the rock glacier is separated from its source area. Its front is about 80 m tall and 40–45° steep, the high angle of repose indicating the presence of ice between the blocks. Rock glacier thickness increases towards the front. The southern lobe consists of brownish weathering schists and paragneiss, shows less pronounced furrows and ridges while its front is less tall and slightly flatter. This lobe receives debris supply from a small cirque and adjacent cliffs rising above its rooting zone. Three small meltwater lakes are located there, exhibiting steep sides and seasonally fluctuating water levels. Water temperatures remain constantly below 1.5°C (fig. 4a; Berger et al., 2004).

Seasonally varying, high deformation rates are recorded in the frontal part, clearly indicating an extensional flow field (up to 1.8 m yr^{-1}; fig. 2.3e,f; Krainer and Mostler, 2006; Hausmann et al., 2012; Groh and Blöthe, 2019). Specifically, high deformation rates recorded during the summer months close to the rock glacier front are attributed to meltwater infiltration into the permafrost body (Krainer and Mostler, 2006), promoting deformation by reducing effective stress (Ikeda et al., 2008).

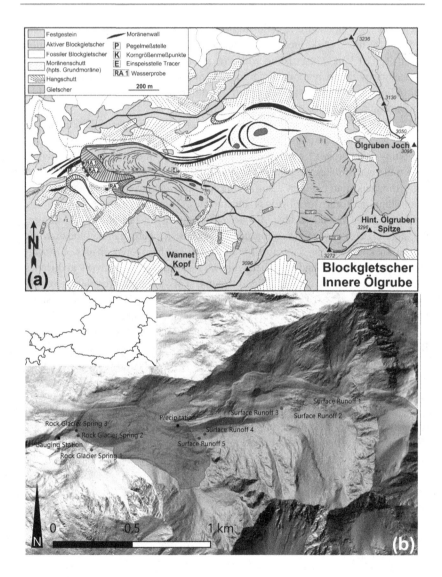

(a) Blockgletscher Innere Ölgrube

(b)

◄**Figure 2.2** (a) Geomorphological map of the Innere Ölgrube (Wagner et al., 2019a). (b) Extent of the rock glacier (red) and its associated catchment (blue) as designated by Wagner et al. (2020a). The gauging station (black square) captures discharge and electrical conductivity of the rock glacier runoff. Electrical conductivity and isotope samples of precipitation (black dot), within the catchment (Surface Runoff 1–5) and at three of the rock glacier springs (Rock Glacier Spring 1–3) are numbered in order of decreasing altitude. (digital elevation model: Open Data Österreich, https://www.data.gv.at, CC BY 4.0; rock glacier inventory: Wagner et al., 2020c, https://doi.pangaea.de/10.1594/PANGAEA.921629, CC BY 4.0; geomorphological-geological map reproduced from "Wasserwirtschaftliche Aspekte von Blockgletschern in Kristallingebieten der Ostalpen. Speicherverhalten, Abflussdynamik und Hydrochemie mit Schwerpunkt Schwermetallbelastungen" by Wagner, T.; Seelig, S.; Wedenig, M.; Pleschberger, R.; Krainer, K.; Kellerer-Pirklbauer, A.; Ribis, M.; Hergarten, S.; Winkler, G., with permission from Gerfried Winkler).

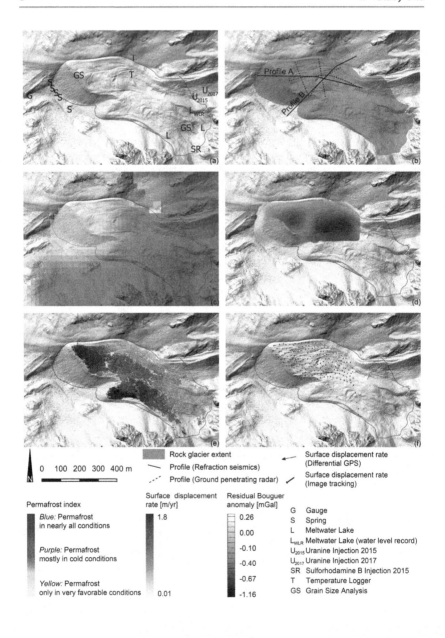

0   100 200 300 400 m

Permafrost index

Blue: Permafrost
in nearly all conditions

Purple: Permafrost
mostly in cold conditions

Yellow: Permafrost
only in very favorable conditions

▨ Rock glacier extent

⌒ Profile (Refraction seismics)

⌁ Profile (Ground penetrating radar)

Surface displacement
rate [m/yr]
1.8

0.01

Residual Bouguer
anomaly [mGal]
0.26

0.00

-0.10

-0.40

-0.67

-1.16

↤ Surface displacement rate
(Differential GPS)

↗ Surface displacement rate
(Image tracking)

G   Gauge
S   Spring
L   Meltwater Lake
L_{WLR} Meltwater Lake (water level record)
U_{2015} Uranine Injection 2015
U_{2017} Uranine Injection 2017
SR   Sulforhodamine B Injection 2015
T   Temperature Logger
GS   Grain Size Analysis

◄**Figure 2.3** Spatial data (a) Rock glacier features and measurement points (b) rock glacier extent (Wagner et al., 2020a,c) and geophysical profiles (reported by Hausmann et al., 2012) (c) permafrost index map (Boeckli et al., 2012a,b) (d) residual Bouguer anomaly (based on Hausmann et al., 2012), (e) surface displacement rates based on image tracking approach (based on Groh and Blöthe, 2019), (f) surface displacement rates based on differential GPS measurements (based on Krainer and Mostler, 2006) (alpine permafrost index map: Boeckli et al., 2021b, https://doi.pangaea.de/10.1594/PANGAEA.784450, CC BY 4.0; digital elevation model: Open Data Österreich, https://www.data.gv.at, CC BY 4.0; rock glacier inventory: Wagner et al., 2020c, https://doi.pangaea.de/10.1594/PANGAEA.921629, CC BY 4.0).

Methods

3

In order to achieve a comprehensive conceptual model of the OEG, the existing information is summarized and complemented by an evaluation of available data using a multi-disciplinary approach. Evaluation of surface displacement rates within the framework of a simple creep model yields a first approximation of the permafrost thickness distribution in those parts of the rock glacier that are not covered by geophysical measurements. A 3D geometrical model of the internal structure is obtained for the complete rock glacier by combining this approximation with the structures identified by the geophysical survey. Analyses of discharge dynamics, water level records, natural and artificial tracers allow for a preliminary estimation of key hydraulic properties and recharge components.

3.1 Data Acquisition and Processing

This section summarizes existing information concerning the OEG and displays available data in a consistent manner.

3.1.1 Geometry

Geophysical measurements (Hausmann et al., 2012) suggest that the rock glacier is built up by a layered internal structure (fig. 2.3, 3.1), including

(1) A coarse grained **active layer** free of ice during the summer months. This layer exhibits an average thickness of 5.2 m, an average total porosity of 0.4, and a bulk density of 1604 kg m^{-3} (Hausmann et al., 2012). Two horizons make up the active layer, an extremely coarse grained surface layer

11

S. Seelig, *Characterizing Groundwater Flow Dynamics and Storage Capacity in an Active Rock Glacier*, BestMasters,
https://doi.org/10.1007/978-3-658-37073-2_3

with blocks up to several meters overlying a finer-grained layer composed of poorly sorted silty sand and sandy silt with embedded boulders (fig. 2.3a; Berger et al., 2004). The > 1 m thick surface layer is composed of variably sized blocks (20–40 cm on average) almost free of fine-grained matrix, and displays strong spatial heterogeneity lacking any distinct pattern. Locally, sharp transitions from coarse-grained to finer grained domains exist (average grain size > 50 cm and < 20 cm, respectively), but gradual transitions occur as well.

(2) An ice-rich **permafrost layer,** exhibiting variable thickness (29 m on average), Ice content (ranging from 31–72 %) and (correspondingly) bulk density (1410–2150 kg m^{-3}). Despite strong heterogeneity, a general trend of increasing thickness and decreasing ice content in downslope direction is observable. Inspection of the longitudinal geophysical profile recorded along the northern rock glacier lobe shows two sections exhibiting increased permafrost layer thickness (Fig. 3.1a). The upper section is associated with a flat bedrock domain, bounded downwards by a steep section resulting in a thinning of the permafrost layer. The increase in thickness close to the rock glacier front is associated with low seismic primary wave velocities and a bedrock slope angle exceeding the surface slope (except at the steep rock glacier front, Fig. 3.1a). Employing a simple creep model, Hausmann et al. (2012) demonstrated that these observations indicate degrading permafrost in this part of the OEG, thus requiring the permafrost layer to thicken in order to keep the rock glacier creeping. Within the northern lobe, permafrost thickness decreases towards the North (Fig. 3.1b).

(3) An ice-free **base layer** below the permanently frozen rock glacier core, consisting of fine-grained sediment, exhibiting a thickness of 10–15 m (average 12 m), a total porosity of 0.3 and a bulk density of 1870 kg m^{-3}.

Note that the porosity values and corresponding bulk densities obtained by Hausmann et al. (2012) are based on seismic compressional wave velocities employing a relationship derived by Watkins et al. (1972). They represent total porosity values, which are expected to exceed the corresponding drainable porosity values, especially in case of the fine-grained base layer.

The underlying bedrock consists of jointed gneiss, located at an average depth of 46 m. A pronounced bedrock threshold in the lower part corresponds to two bedrock outcrops close to the rock glacier margins (fig. 2.3b, 3.1a; Hausmann et al., 2012). It separates a flat bedrock domain above from a distinct steepening of layers below, resulting in a thinning of the permafrost layer (Fig. 3.1a). The

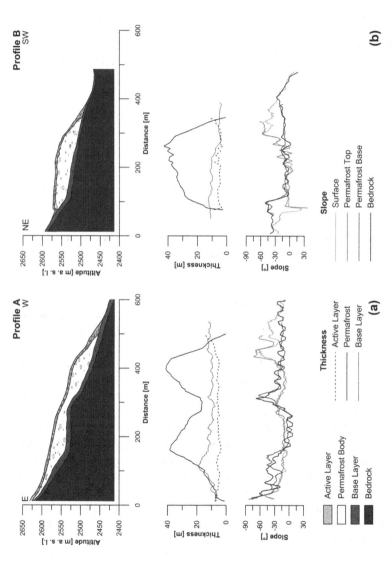

Figure 3.1 Geophysical Profiles combining seismic refraction, GPR, and gravimetry measurements. (a) Longitudinal profile A (b) Transverse profile B. Profile positions are indicated in fig. 2.3b (based on Hausmann et al., 2012)

flat domain is bounded to the north by a pronounced angle in prolongation of a bedrock outcrop (fig. 2.3b, 3.1b; Hausmann et al., 2012).

Sediments surrounding the lower part of the rock glacier mainly comprise talus, subordinately moraine. Seismic refraction measurements indicate 15–20 m of unfrozen sediment covering bedrock (Hausmann et al., 2012), in good agreement with temperature records at the base of the snow cover adjacent to the rock glacier (fig. 2.3a; Berger et al., 2004) and modeled permafrost distribution (fig. 2.3c; Boeckli et al., 2012a,b). However, permafrost is expected to prevail in the sediments surrounding the rock glacier rooting zone (fig. 2.3c).

3.1.2 Hydrology

Five springs at the rock glacier front account for almost the total rock glacier runoff (Berger et al., 2004). Water temperature constantly stays around or below 1°C, indicating water flowing in direct contact with ice within the rock glacier (Krainer and Mostler, 2002; Berger, 2004; Rieder, 2017). A gauging station recording water level and electrical conductivity (EC) at an hourly interval is installed 150 m downstream, after confluence of the creeks emerging from the rock glacier springs. Thus, differentiation of individual springs is not possible. Water level records are converted to discharge values based on a rating curve provided by Heigert (2018), which is based on several discharge measurements employing the salt dilution method during a wide range of flow conditions.

Rock glacier runoff is highly variable, ranging from 6.5 to 718 l/s (Q_{max}/Q_{min} = 110.5). Pronounced sharp (flashy) peaks in response to recharge events are superimposed on base flow characterized by seasonal and diurnal variations (fig. 3.2). The overall hydrodynamic characteristics resemble the superposition of several regularly repeating patterns, exhibiting periods ranging from one day to one year (Heigert, 2018). Winkler et al. (2018b) analyze cyclic patterns of discharge and natural tracers characterizing rock glacier spring flow at diurnal and seasonal time scales. The general discharge pattern is heavily influenced by local weather conditions, with the specific hydraulic response depending on air temperature, solar radiation, precipitation intensity and duration, the timing and magnitude of snow melt events, amount of snow accumulation, and thermal conditions within the active layer (fig. 3.2b; Krainer and Mostler, 2002). Following the onset of snowmelt in early spring, base flow rapidly increases from its minimum, reaching its maximum in early summer. Prominent peaks are observed after heavy rainfall on melting snow cover and intense summer thunderstorms

(Krainer and Mostler, 2002; Heigert, 2018). Temporarily declining air temperatures in summer strongly reduce discharge. Warm, dry weather periods promote diurnal runoff cycles, with time lags between maximum air temperature and peak discharge around 16 h. As summer progresses and snowmelt gradually ceases, peak flow magnitude persistently decreases. Declining air temperatures in autumn rapidly reduce discharge and attenuate the diurnal variations. Following the onset of snowfall, slowly declining base flow dominates that is maintained until snow melt during early spring, occasionally intermitted by single peaks in response to warm rainfall events (Krainer and Mostler, 2002; Heigert, 2018; Wagner et al., 2020b).

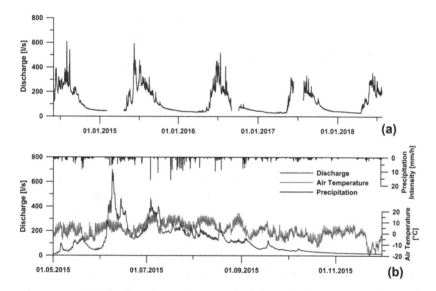

Figure 3.2 (a) Hydrograph of the rock glacier Innere Ölgrube recorded at the gauging station (position indicated in fig. 2.3a). (b) Hydrograph during summer 2015. Note the rapid increase at the beginning of June and flashy response to rainfall events afterwards. Note also the pronounced decline after cold weather characterizing the second half of June and diurnal variations which are clearly observable during dry periods. Peak values gradually decline as summer progresses

Detailed OEG hydrograph analyses are summarized by Wagner et al. (2020b). These authors identify at least two flow components using spring recession analysis, a fast flow component and a slower base flow component, similar to

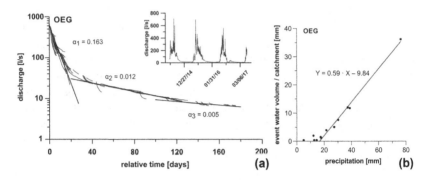

Figure 3.3 (a) Master recession curve analysis and fitted recession exponents (Wagner et al., 2019a; reproduced from "Wasserwirtschaftliche Aspekte von Blockgletschern in Kristallingebieten der Ostalpen. Speicherverhalten, Abflussdynamik und Hydrochemie mit Schwerpunkt Schwermetallbelastungen" by Wagner, T.; Seelig, S.; Wedenig, M.; Pleschberger, R.; Krainer, K.; Kellerer-Pirklbauer, A.; Ribis, M.; Hergarten, S.; Winkler, G., with permission from Gerfried Winkler) (b) Threshold analysis indicating 18 mm of precipitation required to trigger a spring response, followed by 60 % event water flowing quickly towards the spring (Wagner et al., 2020b, https://link.springer.com/article/10.1007 %2Fs00767-020-00455-x, CC BY 4.0)

observations in karst aquifers (fig. 3.3a). By comparing the early recession behavior of active, inactive, and relict rock glaciers, Wagner et al. (2020b) attribute the fast flow component of the (active) OEG to rapid drainage of a perched aquifer on top of the permanently frozen rock glacier core. They relate the slow flow component to the fine grained, unfrozen base layer, stressing the striking similarity between their results and the base layer thickness obtained by the geophysical measurements outlined above: Defining the base flow as runoff during extended periods of very little to no recharge reflected as steady decline in (winter) discharge, and representing the drainage process by a linear storage model, Wagner et al. (2020b) estimate the volume of the associated water volume stored above spring level to encompass between $2.24 \cdot 10^5$ and $5.34 \cdot 10^5$ m^3. Dividing this value by the areal rock glacier extent (0.24 km^2) and assuming a drainable porosity of 0.2, the mean saturated thickness of the corresponding aquifer is estimated to lie between 4.7 and 11.2 m. Wagner et al. (2020b) emphasize that this thickness represents an upper bound estimate based on the assumption that the rock glacier acts as the rate-controlling hydraulic unit of the catchment (i. e. groundwater flux through the rock glacier is similar to or less than through sediments in the higher parts of the watershed).

Employing a threshold analysis, Wagner et al. (2020b) recognize that 18 mm of precipitation (normalized to the OEG catchment area) are necessary to trigger a noticeable increase in spring discharge (fig. 3.3b). Above this threshold they observe a linear correlation between precipitation and event water exhibiting a slope of 0.59. This implies that occasional depressions or pool structures of the bedrock and/or the permafrost table corresponding to a water volume of $\sim 3.3 \cdot 10^4$ m^3 need to be filled before a considerable hydraulic response is initiated at the springs (Wagner et al., 2020b). Slow drainage of these structures might occur via fractures (in permafrost ice and bedrock) or weathering zones (in bedrock) (fill-spill-drain hypothesis). These features promote the hydraulic impact of aquifer base topography on hydraulic conditions within the OEG (either attributed to the surface between two aquifer layers of strongly contrasting hydraulic conductivity, or to bedrock topography). Once this threshold is exceeded ~ 60 % of precipitation flow quickly through the rock glacier, as indicated by the slope in fig. 3.3b (event water; Wagner et al., 2020b).

The alpine, crystalline rock glacier catchment, characterized by short residence time and restricted rock-water interaction allows regarding EC as a conservative tracer (Winkler et al., 2016a). Based on this assumption Heigert (2018) and Wagner et al. (2019b) employed a two component mixing model to differentiate the low mineralized event water component (rainfall, snowmelt, ice melt) from the higher mineralized groundwater component (Fig. 3.4). In general, EC and discharge are inversely correlated. The event water contribution increases during late spring until discharge reaches its maximum in early summer, indicating the maximum level of storage. At this time event water accounts for up to 80 % of the rock glacier discharge. As summer progresses and the overall discharge decreases, the event water share reduces accordingly, resulting in an average contribution around ~ 60 % throughout the summer, in good agreement with the threshold analysis outlined above. Diurnal EC variations mirror the (inversely correlated) discharge variations during dry, warm summer periods (Fig. 3.4, 3.6). After air temperatures falling below the melting point obliterate daily variations in autumn the event water contribution rapidly drops, while groundwater sustains spring flow during the winter (fig. 3.2, 3.4). This transition from the meltwater-dominated period (May–August) to the groundwater-dominated period (September-April) is also reflected in the pronounced EC responses to precipitation events compared to relatively damped response throughout the summer (compare amplitude of discharge response to amplitude of EC response in Fig. 3.4; Heigert, 2018; Wagner et al., 2019b). Time lags between discharge responses and subsequent EC responses to recharge events range from 0 to 12 hours, strongly scattering around a mean of 4 h 20 min (Heigert, 2018).

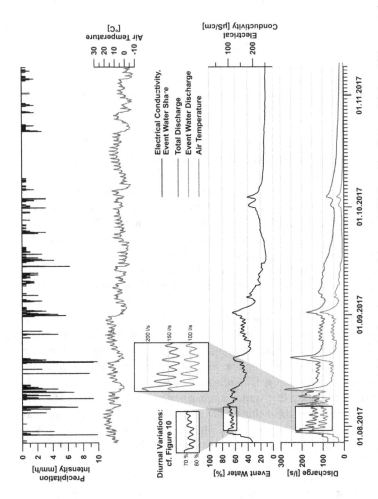

Figure 3.4 Two component mixing model results for summer and autumn 2017 (based on Heigert, 2018, and Wagner et al., 2019b). The highlighted dry period at the beginning of August is analyzed in more detail in fig. 3.6c. Note the changing pattern at the beginning of September. Electrical conductivity and event water share are linearly correlated, thus displayed by one graph with differently scaled axes. See Heigert (2018) and Wagner et al. (2019b) for the complete record

During summer 2017, Heigert (2018) sampled EC of surface runoff within the catchment along a profile extending from the cirque glaciers in the uppermost part down to the rock glacier springs. As depicted in Fig. 3.5 EC generally increases with decreasing altitude, the lowest values being observed at the cirque glacier surface (sample locations are indicated in fig. 2.2b; samples are numbered in order of decreasing altitude). As summer progresses EC increases at each location, with the temporal variation also increasing with decreasing altitude. However, the rock glacier springs (lowest altitude) apparently break these trends, displaying intermediate values as well as a modest overall variation. The increase in EC observed at the catchment runoff samples is attributed to the successively raising contribution from highly mineralized groundwater stored in the talus and moraines that are passed along the flow path (Heigert, 2018). The glacial meltwater share reduces accordingly, with the strongest reduction along the flow path being observed at the beginning of autumn due to restricted meltwater production. Notably, the two lowermost springs (both emerging from the northern rock glacier lobe) exhibit consistently EC values slightly below those recorded in the creek slightly upstream from its infiltration into the rock glacier rooting zone (Surface Runoff 4, 5). This difference increases dramatically in September, now significantly affecting all of the sampled rock glacier springs.

^{2}H and ^{18}O samples taken by Heigert (2018) at the same locations outlined above plot along the Austrian Meteoric Water Line (Hager and Foelsche, 2015), with glacier ice (sample Surface Runoff 1) and precipitation (sampled at the mean catchment altitude) as end members (fig. 3.6a; sample locations are indicated in fig. 2.2b). The catchment samples clearly show the expected progressing depletion in ^{2}H and ^{18}O with increasing altitude (fig. 3.5, 3.6a). The spring samples (lowest altitude) show comparatively little variation, plotting approximately in the center of the catchment sample range. This suggests a common water source of the spring samples, which violates the altitude effect observed in the catchment samples (Heigert, 2018).

A possible explanation consistent with the storage capacity of the rock glacier considers mixing of water from different sources within the rock glacier (Heigert, 2018; Winkler et al., 2018b). Heigert (2018) suggests that melting rock glacier permafrost ice, of low EC and depleted in ^{2}H and ^{18}O, might contribute significantly to spring discharge, acknowledging the uncertainties associated with this conclusion. This deduction is in good agreement with ice samples collected at the nearby rock glacier Inneres Bergli[1], where permafrost ice crops out in the uppermost part of the rock glacier (Heigert, 2018; Winkler et al., 2018b;

---

[1] Exact location given in fig. 5.10 and tab. 5.2

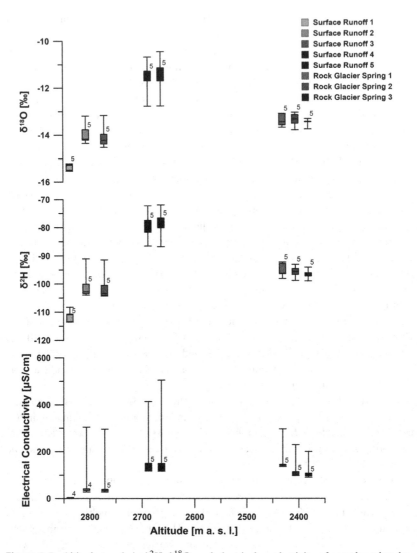

Figure 3.5 Altitude trends in δ^2H, δ^{18}O, and electrical conductivity of samples taken by Heigert (2018) in August and September 2017. Samples collected within the catchment (Surface Runoff 1–5) and rock glacier spring samples (Rock Glacier Spring 1–3) are numbered in order of decreasing altitude. Numbers next to boxplots indicate sample size. Sampling locations are indicated in fig. 2.2b. Note that the spring samples exhibit comparatively low spread and break the altitude trend observable in the surface runoff samples

Wagner et al., 2019b), as well as with the low EC observed at the springs (see above). Note, however, that only a single rainfall sample (collected from 31.7. to 10.8.2017) could be obtained, which represents typically high summer δ^2H and $\delta^{18}O$ values (cf. Clark and Fritz, 1997; Kendall and McDonnel, 1998). Precipitation samples taken during several months at Inneres Bergli indicate δ^2H and $\delta^{18}O$ values spanning the complete record depicted in fig. 3.6a (Winkler et al., 2018b). Snow samples taken there exhibit even larger variability, aside from an overall shift to lower values. The large variance is attributed to the dry inner alpine climate characterized by strong variations in temperature, varying wind exposition and freeze-melt processes within the snow cover (Fritz and Clark, 1997; Kendall and McDonnel, 1998). In addition, it is difficult to differentiate meltwater from glacial and permafrost ice based on their isotopic signature (Winkler et al., 2018b). Wagner et al. (in prep.) assume that glacial meltwater is more important than permafrost ice meltwater, due to the insulating effect of the coarse grained rock glacier surface layer.

Fig. 3.6c depicts diurnal variations in several records observed in early August 2017 (Heigert, 2018; Winkler et al., 2018b). While EC, δ^2H and $\delta^{18}O$ are inversely correlated to discharge, air temperature exhibits a lagged direct correlation suggesting that some kind of melting process causes the daily fluctuations (Heigert, 2018; Winkler et al., 2018b; Wagner et al., 2019a). Plotting EC versus $\delta^{18}O$ reveals a trend towards lower values during that period, superimposed by a diurnal cyclic pattern (fig. 3.6d; Heigert, 2018; Winkler et al., 2018b). At the end of July 2017 the catchment was hit by heavy rainfall (Fig. 3.4) preceding the warm, dry period in early August, suggesting that the pattern reflects changing recharge components: While the general trend indicates a progressive shift from declining rainwater to increasing meltwater contributions, the hysteresis characterizing diurnal cycles indicates a significant impact of a third recharge component (groundwater) affecting spring discharge (Heigert, 2018; Winkler et al., 2018b). This is in perfect agreement with the two component mixing model indicating a groundwater contribution between 30 and 40 % during that period (Fig. 3.4). Based solely on EC, the two component model is actually a projection of the multi component mixing model suggested by Winkler et al. (2018b) for rock glacier recharge, aggregating rainfall, snowmelt, and ice melt into a single component (event water).

A sudden fall in air temperature along with the first snowfall event in early September 2017 fundamentally alters the recharge pattern (Fig. 3.4). Strongly increased EC values indicate an increased groundwater contribution, while meltwater and rainfall components are attenuated (Heigert, 2018). This is supported by isotope trends, where all but the glacier surface sample (Surface Runoff 1)

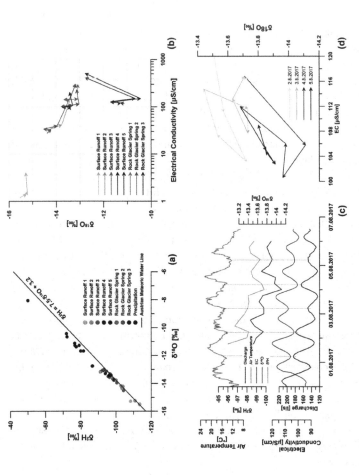

Figure 3.6 (a) Austrian Meteoric Water Line (Hager and Foelsche, 2015) and isotope data (based on Heigert, 2018) (b) δ18O vs. EC at rock glacier springs and at sampling sites in catchment (based on Heigert, 2018) (c) diurnal variations in several parameters at the gauging station during the dry, warm summer period highlighted in Fig. 3.4 (based on Winkler et al., 2018b) (d) diurnal hysteresis with respect to δ18O vs. EC during the dry, warm summer period highlighted in Fig. 3.4 (based on Winkler et al., 2018b). Surface runoff and rock glacier spring samples are numbered in order of decreasing altitude. Sampling locations are indicated in fig. 2.2b

exhibit a convergence of $\delta^{18}O$ values towards -13 ‰ (fig. 3.6b). While this represents a depletion trend for samples Surface Runoff 4 and 5, the remaining samples display an enrichment in ^{18}O relative to ^{16}O compared to their summer values (!), indicating groundwater as common source during autumn (except for Surface Runoff 1; Heigert, 2018; Winkler et al., 2018b).

Wagner et al. (in prep.) analyze the discharge pattern of the OEG catchment by applying a lumped-parameter rainfall-runoff model (GR4J+) that has proven successful in alpine catchment characterization where available data are usually scarce (Wagner et al., 2013, 2016, 2019b). The model includes a snow store as well as an ice store module, both based on a degree-day factor where ice melt is permitted while snow cover is absent (i. e. completely drained snow store). The obtained results are plotted in Fig. 3.7 along with the accordingly calculated recharge components providing the model input. The calibrated model indicates a relatively large storage component which is attributed to the unfrozen base layer of the OEG (Wagner et al., in prep.). Conversely, the catchment provides restricted soil moisture storage only, resulting in relatively low evapotranspiration rates. This is in good agreement with the predominance of bare rocks and the absence of thick soil cover in the catchment as outlined above. The relative contributions of rainfall, snowmelt, and ice melt as depicted in Fig. 3.7 account for 33 %, 37 %, and 30 %, respectively (Wagner et al., 2020b). These results are in good agreement with the relative contributions obtained from analyses of EC and stable isotope signatures.

3.2 Internal Structure

This section develops an approach that combines several methods to obtain a 3D geometrical model of the OEG internal structure. The resulting geometrical model consists of a set of digital elevation models specifying the elevation of bedrock, permafrost base, permafrost top, and rock glacier surface throughout the OEG. These interfaces form the boundaries between major hydrostratigraphic units and can easily be integrated in a prospective numerical groundwater flow model.

The thickness of the active layer (AL), being defined by thermal constraints, varies throughout the year as well as with local variations in relief and grain size. Increasing thickness during the summer months followed by decreasing thickness in autumn is caused by melting and refreezing of interstitial ice forming the uppermost part of the frozen rock glacier core. Generally, the AL is well developed along ridges but poorly along furrows, due to locally delayed snow melt

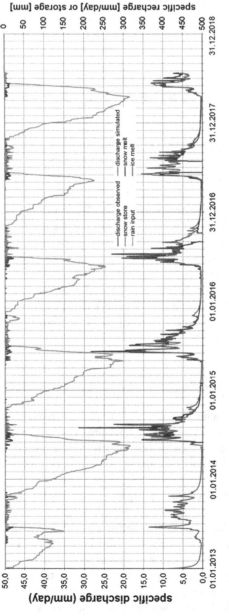

Figure 3.7 Observed (blue) and modeled (red) specific discharge of the Innere Ölgrube catchment based on the calibrated GR4J + rainfall-runoff model (Wagner et al., in prep.). The snow store implemented in the model approximates the mean snow cover water equivalent throughout the catchment (grey). Recharge components are differentiated into rainfall, snow melt, and ice melt showing seasonal trends regarding their relative contributions (based on Wagner et al., in prep.)

(Fig. 3.8) and geometrical constraints affecting heat transport. Domains characterized by extremely coarse grained material exhibit reduced AL thickness due to the cooling influence of openwork blocky debris (Barsch, 1996; Wagner et al., 2019b). On the long term, AL thickness is expected to increase due to degrading permafrost in the Alps and rapid downslope movement of the OEG, continuously changing its position relative to the lower permafrost boundary.

Seismic refraction measurements identifying the AL base along two profiles were carried out in summer 2003 (Hausmann et al., 2012). Uncertainties related to spatial and temporal variation of AL thickness as well as unknown changes having potentially occurred during the 17 years since the measurements were carried out do not justify a detailed, high-resolution specification of the AL base elevation. However, since the permeability of the AL is expected to exceed the permeability of the underlying permafrost body, probably by several orders of magnitude, the AL base defines the bottom of a shallow, highly permeable aquifer within the OEG. As such, its shape is expected to exert a controlling influence on groundwater conditions within this perched aquifer. Regarding a prospective ground water flow model, the uncertainty in AL thickness plays only a subordinate role compared to the uncertainty in hydraulic conductivity of the AL and the underlying permafrost layer. Thus, balancing representation of the most important features of the AL against data scarcity, the following strategy is proposed: Since a high-resolution (1 m) digital elevation model of the rock glacier surface is available, this surface is smoothed employing the cubic convolution resampling algorithm provided by ArcGIS 10.4 (ESRI, 2016). Lowering the surface obtained in this way by the mean AL thickness (5.2 m) yields an approximation of the AL base. In this way its micro relief, which is believed to strongly affect hydrologic processes within the AL (Burger et al., 1999; Krainer and Mostler, 2002; Berger et al., 2004; Jones et al., 2019) is represented conceptually. The smoothing takes account of the diffuse 3D nature of heat transport within the AL including differential thermal boundary conditions (e. g. snow in furrows) by increasing AL thickness below ridges while reducing it along furrows (Fig. 3.8). However, this approach fails to represent features such as crevasses, taliks or channels eroded into the permanently frozen rock glacier core, which are believed to play an important role in active rock glacier hydrology (Burger et al., 1999; Arenson et al., 2010; Jones et al., 2019).

Combined analysis of ground penetrating radar (GPR), refraction seismics, and gravimetric measurements evaluated by Hausmann et al. (2012) allow a detailed assessment of the deeper internal rock glacier structure. The elevation of the permafrost base derived from the GPR and seismic profiles (fig. 2.3b) provides control points for ordinary kriging implemented using ArcGIS 10.4. The permafrost thickness calculated from the interpolated interfaces is compared to the residual Bouguer anomaly yielding additional information on density and ice content (fig. 2.3d).

Little information is available concerning the geometry of the base layer. The distance between the geophysical profiles is large compared to its thickness, severely limiting any interpolation approach. However, its average thickness of 12 m indicated by these profiles agrees well with the saturated aquifer estimates obtained from winter recession analysis of spring flow (Wagner et al., 2020b). Therefore, as a first approximation, a constant thickness of 12 m is assumed throughout the rock glacier.

Figure 3.8 Snow covering the furrows and depressions of the rock glacier Innere Ölgrube on 2.6.2020, while ridges are already free from snow (Photos: Andreas Pilz)

As evident from fig. 2.3b geophysical measurements cover the lower part of the rock glacier (approximately 50 %). Surface topography and long-term surface displacement rates (fig. 2.3e), representing average values from 2003 to 2015, are the only information available for characterizing the rock glacier rooting zone (Groh and Blöthe, 2019). Therefore, the rest of this section outlines a method for deriving an estimate of permafrost thickness based on these surface displacement rates, surface slope angles calculated from the digital elevation model, AL thickness estimated according to the procedure outlined above, and mean values of AL and permafrost layer bulk density provided by Hausmann et al. (2012).

The fundamental process driving active rock glacier deformation is continuous creep of stress transferring ground ice (Kääb et al., 2003; Haeberli et al., 2006). Assuming deformation to be continuously distributed throughout the permanently frozen rock glacier core allows for an estimation of permafrost thickness from surface displacement rates. Provided the active layer rides passively on top of a frozen rock glacier core deforming according to a power-type creep law, integrating the displacement rate profile depicted in fig. 3.9a yields the surface displacement rate (Nye, 1957; Wahrhaftig and Cox, 1959; Haeberli, 1985; Whalley and Martin, 1992; Barsch, 1996; Konrad et al., 1999; Cuffey and Paterson, 2010; Hausmann et al., 2007). Conversely, the thickness of the deforming permafrost layer might be derived from measured surface displacement rates by reversing the calculation procedure (Konrad et al., 1999; Hartl et al., 2016). Since this method is based on clearly defined mechanical principles and accounts for site-specific rock glacier characteristics (slope, displacement rate, bulk density), the obtained results are expected to be more accurate than those obtained from more general (empirical) approaches (Jones et al., 2019).

However, the actual creep behavior of rock glaciers is not fully captured by a simple creep model and the rheology of the ice-debris mixtures is generally poorly defined (Haeberli et al., 2006). Uncertainty arises from the fact that active rock glaciers typically exhibit distinct shear horizons, where most of the deformation takes place (Barsch, 1996; Haeberli et al., 1998, 2006; Arenson et al., 2002; Buchli et al., 2013, 2018; Krainer et al., 2015). In addition, deformation patterns are affected by the highly heterogeneous structure of rock glaciers, topography effects (e. g. shallow bedrock, change in slope, irregular bedrock surface), infiltrating water causing seasonal variations in displacement rates and additional failure mechanisms causing abrupt slippage along failure surfaces and distinct periods of movement (Haeberli, 1985; Giardino and Vitek, 1988; Barsch, 1996; Arenson et al., 2002; Żurawek, 2002; Haeberli et al., 2006; Jansen and Hergarten, 2006; Ikeda et al., 2008; Buchli et al., 2013, 2018). As a result, differences in surface displacement rates among individual rock glaciers currently cannot be unequivocally attributed to differences in thickness, internal structure, topography or climate (Kääb et al., 2003; Haeberli et al., 2006). The analysis presented below focuses on the long-term creep behavior of the OEG, based on surface displacement rates averaged over 12 years. Consequently, any temporal variations in deformation rates during that period are indistinguishable from steady (secondary) creep rates and integrated in the model (cf. Jansen and Hergarten, 2006). In other words, short-time or periodical deformation rates caused by additional deformation processes are attributed to secondary creep, resulting in a biased estimate.

Despite these uncertainties the employed model has proven successful in predicting permafrost thickness at two active rock glaciers: Konrad et al. (1999) compared model predictions to core drillings at the Galena Creek Rock Glacier (Absaroka Mountains, Wyoming). Hartl et al. (2016) verified their results using GPR measurements at the Äußeres Hochebenkar Rock Glacier (Ötztal Alps, Austria)[2]. In both cases the model yielded acceptable results. Within a single rock glacier, surface displacement rates are generally correlated with surface slope (Konrad et al., 1999; Kääb et al., 2003; Haeberli et al., 2006). Therefore, while derived absolute values should be treated with care, the model is likely to represent the overall geometry of the permanently frozen rock glacier core appropriately.

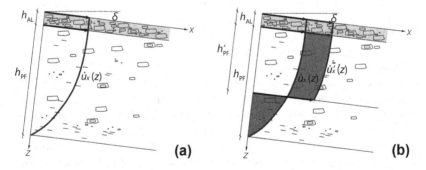

Figure 3.9 (a) Creep model prediction of differential displacement rate. Integrating the depicted profile yields the surface displacement rate. (b) Potential pitfall arising from the simplified creep mechanism: The surface displacement rate (area beneath the respective profile) caused by distributed creep throughout a permafrost body of thickness h_{PF} equals the surface displacement rate raised by a thinner (h_{PF}') permafrost body exhibiting concentrated deformation along a distinct shear zone (red colored areas are equal); c. f. Nye (1957) for a similar glacier model

---

[2] See fig. 5.10 and tab. 5.2 for the exact location of these rock glaciers.

Consider a right-handed Cartesian coordinate system, the x-axis pointing in dip direction along the rock glacier surface, the y-axis pointing in strike direction, and the z-axis pointing downwards toward the rock glacier base (fig. 3.9). Scalars are given in italics, vectors in bold italics, tensors bold, tensor components are indicated by subscripted indices. The ice-debris mixture of the permanently frozen rock glacier core deforms in response to the rock glacier's weight, while the overlying coarse debris layer (active layer, free of ice) is transported passively on top. The fine grained, unfrozen base layer underlying the permafrost core is not incorporated into the model. As a first-order approximation, assume the ice-rock mixture to be incompressible and mechanically isotropic. The assumption of incompressibility is considered appropriate for long-term descriptions of rock glacier deformation (Kääb et al., 2003), although as a very rough approximation only (cf. Haeberli et al., 2006). Although clearly a severe simplification, neglecting anisotropy is regarded acceptable, given that the permafrost thickness is large compared to the dimension of single irregularities such as ice lenses (Haeberli, 1985).

Assume that changes in displacement rate along the y-axis are small (plane strain assumption). The thickness of the permafrost core is small compared to its length and width, thus the deforming system might be represented by a parallel-sided slab of ice-debris mixture thickness h_{PF} [m] and constant bulk density ρ_{PF} [kg m^{-3}], carrying an active layer of thickness h_{AL} [m] and bulk density ρ_{AL} [kg m^{-3}]. The slab is resting on a rough plane of slope angle δ [−] corresponding to the permafrost base (fig. 3.9a). Due to its weight, a column of unit cross section oriented perpendicular to the permafrost table, exerts a driving stress on any (hypothetical) parallel plane within the frozen rock glacier core given by

$$\sigma_{xz} = (\rho_{PF}(z - h_{AL}) + \rho_{AL}h_{AL})g \sin \delta \qquad \text{(Eq. 3.1)}$$

(gravitational acceleration g [m s^{-2}], stress σ [Pa]). Eq. 3.1 approximately holds for wedge-shaped slabs too, provided base and surface slope angles are small (roughly < 20°) (cf. Cuffey and Paterson, 2010). However, longitudinal stress gradients are neglected, restricting its application to sufficiently large scales (i. e., averaged values of h, δ, and ρ). Differentiating eq. 3.1 with respect to z shows that σ_{xz} increases linearly with depth

$$\sigma_{xz} = \sigma_1 + \frac{z - h_{AL}}{h_{PF}}(\sigma_2 - \sigma_1) \qquad \text{(Eq. 3.2)}$$

where σ_1 and σ_2 correspond to the driving stress at the permafrost top and base, respectively (given by evaluating eq. 3.1 at $z_1 = h_{AL}$ and $z_2 = h_{AL} + h_{PF}$). Define an effective deviatoric stress τ_e [Pa] and an effective strain rate $\dot{e}_e$ [s^{-1}] according to (Nye, 1952, 1953, 1957; Arenson and Springman, 2005)

$$\tau_e^2 = \frac{1}{2}\left(\tau_{xx}^2 + \tau_{yy}^2 + \tau_{zz}^2\right) + \tau_{xy}^2 + \tau_{xz}^2 + \tau_{yz}^2 \qquad \text{(Eq. 3.3)}$$

$$\dot{e}_e^2 = \frac{1}{2}\left(\dot{e}_{xx}^2 + \dot{e}_{yy}^2 + \dot{e}_{zz}^2\right) + \dot{e}_{xy}^2 + \dot{e}_{xz}^2 + \dot{e}_{yz}^2 \qquad \text{(Eq. 3.4)}$$

(deviatoric stress τ [Pa], strain rate $\dot{e}$ [s^{-1}]). Thus, τ_e^2 and $\dot{e}_e^2$ correspond to the second invariant of the deviatoric stress and strain rate tensor, respectively (i. e., the second coefficients of their respective characteristic polynomials). Note that with respect to eq. 3.4 this holds for incompressible material only (e. g. Drucker, 1967).

Assume each strain rate component to be proportional to its corresponding deviatoric stress component (Nye, 1957)

$$\dot{e}_{ij} = \psi_1 \tau_{ij} \qquad \text{(Eq. 3.5)}$$

(proportionality constant ψ_1 [Pa^{-1} s^{-1}]). The deformation of frozen debris is attributed to plastic deformation of pore ice, readjustment of debris particles, pressure melting of ice at grain contacts, breakdown of ice and structural bonds to the grains, as well as migration of unfrozen water present within the permanently frozen rock glacier core (Haeberli, 1985). The analysis focuses on the long term creep behavior, thus elastic deformation might be neglected and the rock glacier assumed to be in a permanent state of steady (secondary) creep (Haeberli, 1985). Although a general constitutive model for permafrost in alpine rock glaciers has not been established yet (Haeberli et al., 2006), τ_e and $\dot{e}_e$ are commonly assumed to be related by a simple power law (Glen, 1955; Nye, 1957; Haeberli, 1985)

$$\dot{e}_e = A\,\tau_e^{\,n} \qquad \text{(Eq. 3.6)}$$

The creep parameter A [Pa^{-n} s^{-1}] is a function of temperature, volumetric ice content, grain size, and fabric (Arenson and Springman, 2005; Haeberli et al., 2006). The creep exponent n [$-$] is a function of volumetric ice content (Arenson and Springman, 2005). To keep the model simple and comparable to results obtained by Konrad et al. (1999) and Hartl et al. (2016), constant values of $n = 3$, $A = 2.4 \cdot 10^{-24}\,\mathrm{Pa^{-3}\,s^{-1}}$ are applied. Note, however, that the hetero-geneous internal structure of active rock glaciers and the seasonally varying boundary conditions actually imply significant spatial and temporal variations in both parameters (Haeberli et al., 2006). Combining eq. 3.3–3.6 yields

$$\dot{e}_{jk} = A\,\tau_e^{\,n-1}\,\tau_{jk} \qquad\qquad (\text{Eq. 3.7})$$

In general, active rock glaciers are characterized by zones of longitudinal compression and extension (Kääb et al., 2003; Haeberli et al., 2006). The employed model neglects longitudinal strain rates in order to simplify the calculation. The ice-debris mixture responds to this situation by deforming in simple shear, implying that the only nonzero strain rate component is given by $\dot{e}_{xz} = (\partial \dot{u}_x/\partial z)/2$ ($\dot{u}_z = 0$, flow lines are parallel), where $\dot{u}$ [m s^{-1}] denotes displacement rate. Using eq. 3.7

$$\frac{\partial \dot{u}_x}{\partial z} = 2A\,\tau_e^{\,n-1}\,\sigma_{xz} \qquad\qquad (\text{Eq. 3.8})$$

Note that in such a situation eq. 3.7 predicts $\tau_{jk} = 0$ except for τ_{xz}, since $\dot{e}_{jk} = 0$ except for $\dot{e}_{xz}$, A and τ_e are $\neq 0$ throughout the permanently frozen rock glacier core. Therefore, using eq. 3.3, eq. 3.8 reduces to

$$\frac{\partial \dot{u}_x}{\partial z} = 2A\,\sigma_{xz}^{\,n} \qquad\qquad (\text{Eq. 3.9})$$

However, the model could easily be extended to incorporate compressive and extending deformation regimes by relieving this rather strict assumption, as demonstrated by Nye (1957) with respect to glacier mechanics (cf. Cuffey and Paterson, 2010).

Substituting from eq. 3.2 and integrating eq. 3.9 yields the surface displacement rate $\dot{u}_{x,s}$

$$\dot{u}_{x,s} = \dot{u}_{x,b} + \frac{2A}{(n+1)}(\rho_{PF}\,g\sin\delta)^n\left(\left(h_{PF} + \frac{\rho_{AL}}{\rho_{PF}}h_{AL}\right)^{n+1} - \left(\frac{\rho_{AL}}{\rho_{PF}}h_{AL}\right)^{n+1}\right)$$

$$(\text{Eq. 3.10})$$

where $\dot{u}_{x,b}$ denotes the displacement rate at the permafrost base. Assuming the latter to be negligible, eq. 3.10 corresponds to eq. 3.4 by Konrad and Humphrey (2000), and rearranging yields

$$h_{PF} = \left(\frac{\dot{u}_{x,s}(n+1)}{2A(\rho_{PF}g\sin\alpha)^n} + \left(\frac{\rho_{AL}}{\rho_{PF}}h_{AL} \right)^{n+1} \right)^{\frac{1}{n+1}} - \frac{\rho_{AL}}{\rho_{PF}}h_{AL} \qquad \text{(Eq. 3.11)}$$

This is the equation given by Konrad et al. (1999) and Hartl et al. (2016), corrected for two typographic errors.

Note that the simplifications inherent in the simple creep model give erroneous results in case the outlined assumptions are violated. A possible pitfall is indicated in fig. 3.9b: According to eq. 3.11, an observed surface displacement rate $\dot{u}_{x,s}$ corresponds to a permafrost thickness h_{PF}, assuming the permanently frozen rock glacier core exhibits distributed secondary creep. However, the same surface displacement rate might be obtained from a thinner permafrost core in case a significant part of the deformation is localized along a narrow shear zone (note that the red areas in fig. 3.9b are equal).

Surface slope angle and AL thickness are averaged over domains of 25 × 25 m employing the same cubic resampling algorithm used above (ArcGIS 10.4). Based on these average values, permafrost thickness is estimated using eq. 3.11. Subtracting the thickness derived in this way from the AL base (approximated according to the procedure outlined above) yields the permafrost base elevation. Subtracting another 12 m (thickness of base layer, assumed to be constant) gives the bedrock elevation. As a result, the surfaces bounding the three major hydrogeological units of the OEG (active layer, permafrost layer, and base layer) are continuously specified throughout the rock glacier (bedrock, permafrost base, permafrost top, rock glacier surface).

3.3   Recession Analysis

This section outlines a procedure to estimate the base layer's hydraulic conductivity from the long-term recession behavior of the rock glacier springs. The creeks emerging from these springs confluence upstream of the gauge, while slightly downstream of the gauge the creek bed is formed by outcropping bedrock (fig. 2.2a). Consequently nearly all of the groundwater emerging from the rock glacier is forced to resurface at the spring group or at least upstream from the gauge (Berger et al., 2004). Therefore, the recorded spring flow carries an imprint of the entire catchment, enabling the derivation of aquifer properties from the hydrograph (Rehrl and Birk, 2010).

Assume the spring discharge Q [m^3 s^{-1}] is given by a superposition of J aquifer components aligned in parallel. The total discharge is given by the sum of all draining aquifer component contributions Q_j [m^3 s^{-1}] (Fig. 3.10a). The exact contribution of each component is a priori unknown, as indicated by various possible realizations (grey lines) in Fig. 3.10a (depicting the master recession curve analyzed by Wagner et al., 2020b; a similar sketch can be drawn for single event analysis). During periods of negligible recharge, the spring flow from each aquifer component is given independently from its detailed geometry, internal structure, or boundary conditions by an infinite series of exponential functions (Sahuquillo, 1983). However, most of the summands approach zero within a short period. The length of this period is influenced by the aquifer geometry as well as by the initial conditions (Birk and Hergarten, 2010). Thus, after sufficiently long time has elapsed, the total spring flow might be approximated by

$$Q = \sum_{j=1}^{J} \sum_{i=1}^{\infty} B_{ij} e^{-\alpha_{ij} t} \approx \sum_{j=1}^{J} B_{1j} e^{-\alpha_{1j} t} \qquad \text{(Eq. 3.12)}$$

(discharge contribution B_{ij} [m^3 s^{-1}] at the beginning of the recession, time t [s], recession coefficient α_{ij} [s^{-1}]). The total discharge at the beginning of the recession is given by $Q_0 = \Sigma B_{ij}$. According to the geometrical model outlined above, the aquifer components of the spring discharge can be identified with contributions from the base layer, permafrost layer, and active layer, respectively ($J = 3$). Due to the restricted storage capacities of the latter two, all but the first will run dry during extended periods of little to no recharge. Specifically, the base layer presumably sustains spring flow during the lengthy alpine winter periods characterized by thick snow cover and low temperatures (Wagner et al., 2020b). As winter advances, the rock glacier spring recession is increasingly better approximated by a single exponential function (Maillet, 1905; Kresic and Stevanovic, 2010). This approximation is plotted as black line in Fig. 3.10a.

As outlined above, Wagner et al. (2020b) constrained the corresponding parameters (B_{11} and α_{11}) by employing least squares regression accounting for the observed long-term recession only. By integrating the extrapolated function and normalizing the obtained volume to the rock glacier area, they derived an estimate of the saturated thickness b [m] at the beginning of the recession period. As shown by Harrington et al. (2018) for the (inactive) Helen Creek Rock Glacier

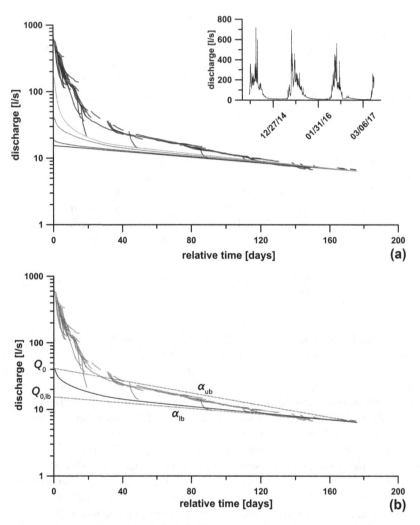

Figure 3.10 Recession analysis (a) different realizations of the base layer contribution (grey lines) and linear storage approximation (black) (b) sketch indicating estimated K values are actually lower bound estimates (modified after Winkler et al., 2018b). The solid line indicates the (unknown) true recession behavior, while the dotted lines indicate upper and lower bound estimates based on the linear storage model (with upper and lower bound recession coefficients α_{ub} and α_{lb}, respectively). The discharge at the beginning of the recession Q_0 is underestimated by the lower bound estimate $Q_{0,lb}$, but equals the (hypothetical) upper bound estimate (see text)

(Canadian Rockies, Alberta)[3], this allows for an estimate of the hydraulic conductivity based on Darcy's Law at the beginning of the winter recession (Darcy, 1856)

$$K = -\frac{Q}{bw}\frac{\Delta l}{\Delta \varphi} \qquad \text{(Eq. 3.13)}$$

(hydraulic conductivity K [m s^{-1}], length Δl [m], aquifer width w [m], hydraulic head difference $\Delta \varphi$ [m]).

However, the obtained estimate is biased because the additional terms outlined on the left hand side of eq. 3.12 cannot be neglected at the beginning of the recession (up to at least 100 days ($= 0.5/\alpha_{11}$), cf. Birk and Hergarten, 2010; after 100 days, however, eq. 3.13 cannot be evaluated because b is certainly not uniform, cf. Pauritsch et al., 2015). Eq. 3.13 actually yields a lower bound estimate of K. To see this, consider the sketch provided by Fig. 3.10b. The black solid line indicates the (unknown) true base layer contribution to spring flow. The lower dotted line corresponds to the linear storage approximation employed by Wagner et al. (2020b). Integration of the former gives the true base layer water volume stored at the beginning of the recession period V_0 [m^3], while integration of the latter gives a lower bound estimate $V_{0,lb}$ ($V_{0,lb} < V_0$, yielding $b_{0,lb} < b_0$ by assuming a uniform saturated thickness at the beginning of the recession). Note that $Q_{0,lb}$ as well as $b_{0,lb}$ underestimate the actual values, precluding a statement about their relative impact on the estimated hydraulic conductivity, K_{est} (eq. 3.13), at this stage of the analysis. To solve this problem, a (hypothetical) upper bound estimate based on the linear storage model is indicated by the upper dotted line in Fig. 3.10b (note $\alpha_{lb} < \alpha_{ub}$ and $Q_0 = Q_{0,ub}$). As evident from the sketch, $V_{0,lb} < V_0 < V_{0,ub}$, where $V_{0,lb} = Q_{0,lb}/\alpha_{lb}$ and $V_{0,ub} = Q_0/\alpha_{ub}$. Expressing the stored water volumes by their respective saturated thicknesses shows

$$\frac{b_0}{b_{0,lb}} < \frac{Q_0}{Q_{0,lb}}\frac{\alpha_{lb}}{\alpha_{ub}} \qquad \text{(Eq. 3.14)}$$

The estimate K_{lb} of the true hydraulic conductivity K is actually a lower bound estimate if $K_{lb}/K < 1$. In this case, according to eq. 3.13, $(Q_{lb}/b_{lb})/(Q/b) < 1$. Evaluating Q and b at the beginning of the recession period and rearranging shows

---

[3] Exact location given in fig. 5.10 and tab. 5.2

$$\frac{b_0}{b_{0,lb}} < \frac{Q_0}{Q_{0,lb}} \qquad\qquad\text{(Eq. 3.15)}$$

Eq. 3.14 satisfies eq. 3.15 regardless of the exact functional relationship describing the true base layer contribution since $\alpha_{lb} < \alpha_{ub}$ in any case (see sketch). Thus, estimating the hydraulic conductivity of the base layer based on the recession analysis provided by Wagner et al. (2020b) poses a lower bound on the actual hydraulic conductivity. In case the flow geometry and initial conditions within the base layer were known, an improved estimate could be derived (e. g. Brutsaert, 1994; Kovács et al., 2005; Hergarten and Birk, 2007; Birk and Hergarten, 2010; Winkler et al., 2016). As emphasized by Wagner et al. (2020b), and additional error of unknown magnitude arises from the fact that storage in the catchment (upstream from the rock glacier) is neglected, resulting in an upper bound estimate for b (lower bound estimate for K). In case catchment hydraulic characteristics were known, an improved estimate could be derived.

3.4   Thermokarst Lake Water Level Fluctuations

This section roughly estimates the hydraulic conductivity of the permanently frozen rock glacier core. Thermokarst lakes drain either via diffuse infiltration into the permafrost body, into thermokarst features (crevasses, taliks) or via overflow followed by downstream runoff along the permafrost table. Ice-rich ground and low relief promote the development of thermokarst lakes, where the pooling of water in depressions formed by thaw settlement further thaws the permafrost beneath (Kääb and Haeberli, 2001). The formation of these features requires the presence of impermeable permafrost preventing rapid percolation into the ground, ice-supersaturation as precondition for thaw settlement and sufficient flat topography to inhibit rapid melting of topographic barriers damming the lake (Kääb and Haeberli, 2001). This suggests that the permafrost layer is continuous and rather impermeable in the upper part of the rock glacier where thermokarst lakes are present (fig. 2.3a). Water level fluctuations in the northernmost meltwater lake located in the rock glacier rooting zone were recorded at an interval of 30 min between 15.7.2015 and 7.9.2016 (fig. 2.3a, 3.11–3.13). The lake is interpreted as thermokarst feature by Berger et al. (2004), suggesting that the substrate at its base is sealed by permafrost ice. The constantly low water temperature recorded by Berger et al. (2004) and the likelihood of permafrost occurrence modeled by Boeckli et al. (2012a,b) support this conclusion (fig. 2.3c).

Figure 3.11   Surveyed thermokarst lake. Its position is indicated in fig. 2.3a, recorded water level fluctuations are depicted in Fig. 3.12 and 3.13 (Photos: Thomas Wagner)

The observed water level fluctuations are used to obtain an estimate of the hydraulic conductivity at the bottom of the lake. After the cessation of snowmelt, water levels during dry periods remain remarkably stable once the recession from preceding precipitation events has flattened out (Fig. 3.13). During the observation period, six of these periods can be identified by approximating the end of snowmelt using the snow store of the calibrated rainfall-runoff model (Wagner et al., in prep.). Assume that during these periods, drainage of the lake does not occur via distinct outlets but only by diffuse infiltration into the underlying saturated permafrost body. For lake water infiltrating distributed across the lake bottom vertically downward (hydraulic gradient of ~ 1), the specific discharge corresponds to the lake water level reduction. Thus, dividing the water level drop during a dry period by its duration yields the hydraulic conductivity of the frozen rock glacier core underlying the thermokarst lake. However, the hydraulic conductivity obtained in this way is actually an upper bound estimate, since no attempt is made to quantify the evaporation rate from the lake surface (necessary level of accuracy difficult to achieve).

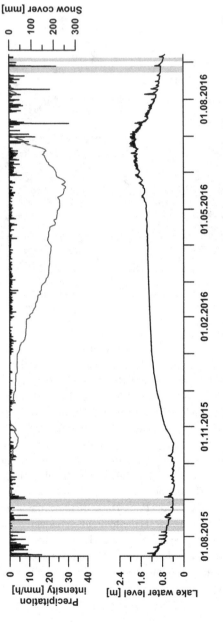

Figure 3.12 Thermokarst lake water level recorded between 15.7.2015 and 7.9.2016. Precipitation and specific snow cover storage (based on Wagner et al., in prep.) are included to identify dry periods after the cessation of snow melt. The six evaluation periods depicted in Fig. 3.13 are indicated by grey bars. The smooth winter curve might be associated with freezing of the pond's surface

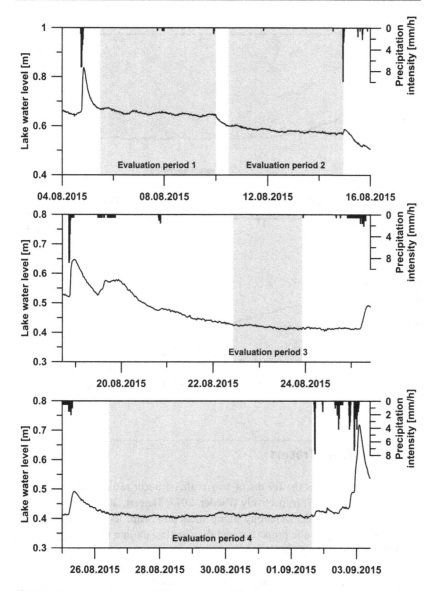

Figure 3.13 Evaluation periods for water level fluctuations of the surveyed thermokarst lake (cf. Fig. 3.12). Once the recession following rainfall events levels off, dry periods of relatively constant water level are evaluated to estimate the infiltration rate. In all depicted cases, snow melt had already ceased as indicated by the results of the calibrated rainfall runoff model (Wagner et al., in prep.), depicted in Fig. 3.7 and 5.1

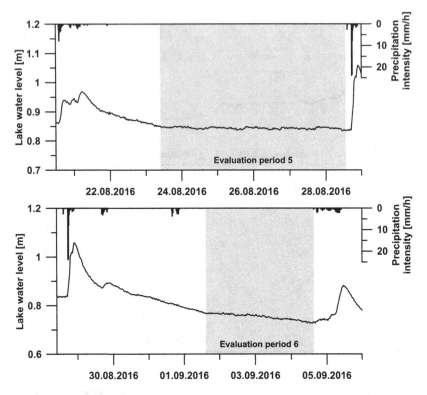

Figure 3.13 (continued)

3.5   Artificial Tracers

This section reanalyzes the results of two artificial tracer tests conducted at the OEG in 2015 and 2017, respectively (Rieder, 2017; Heigert, 2018). Since tracers integrate the small-scale variability along their flow path, they are well suited for inferring key hydraulic properties and drawing insight into the flow processes taking place within the rock glacier.

The first tracer test was implemented in 2015 using uranine (C. I. 45 350, $C_{20}H_{10}Na_2O_5$) and sulforhodamine B (C. I. 45 100, $C_{27}H_{29}N_2NaO_7S_2$) as tracers. Uranine exhibits a solubility of 300 g l^{-1} (at 20°C water temperature), its fluorescence intensity is slightly dependent on water temperature. It is hardly affected by sorption and degrades irreversibly in sunlight. It exhibits a theoretical detection limit of 10^{-3} μg l^{-1} in clear water, while under field conditions a detection limit of $2 \cdot 10^{-2}$ μg l^{-1} is achievable (Schnegg, 2002; Käss, 2004; Leibundgut et al., 2009). Sulforhodamine B exhibits a solubility of 10 g l^{-1} (at 10°C water temperature), a theoretical detection limit of. $3 \cdot 10^{-2}$ μg l^{-1} in clear water, but 0.2 μg l^{-1} under field conditions (Schnegg, 2002). Its fluorescence intensity is strongly dependent on water temperature, it is slightly susceptible to sunlight and slightly more prone to sorption than uranine (Käss, 2004; Leibundgut et al., 2009).

On 15.7.2015, 25 g of uranine previously dissolved in 1 l water were injected at 11:20 into a small creek slightly upstream from its infiltration into the rock glacier rooting zone (fig. 2.3a). Subsequently 25 g of Sulforhodamine B were injected into the rooting zone of the southern lobe (fig. 2.3a). Fluorescence intensity was measured every minute at the gauging station, 980 m (linear distance) from the uranine injection point and 930 m from the sulforhodamine B injection point, using a field fluorometer GGUN FL30 (Schnegg, 2002) from 15.7.2015 08:52 to 21.7.2015 12:53. The average water temperature at the gauging station was 2.7°C during that period. In 2017, 202.49 g of uranine previously dissolved in 2 l water were injected at 17:08, 31.7.2017, slightly downstream from the injection site in 2015 (fig. 2.3a). The field fluorometer was again installed at the gauging station, at a linear distance of 910 m from the injection site, recording fluorescence intensity every 2 min at an average water temperature of 2.8°C. Water samples were taken for subsequent fluorometer calibration at the University of Graz (average water temperature during calibration 26.3°C (2015) and 22.0°C (2017), respectively).

Fluorescence intensity decreases with increasing temperature, the exact rate of decrease depending on tracer composition. Due to the strong water temperature differences between field test and calibration, recorded signals are corrected for temperature according to Smart and Laidlaw (1977), as recommended by Käss (2004) and Leibundgut et al. (2009). The background noise level induced by absorption and scattering by suspended particles, air bubbles or auto fluorescence is defined as arithmetic mean of the records taken before tracer injection and subtracted from the time series (Field, 2002). Discharge and tracer concentration measurement intervals are harmonized employing spline interpolation of discharge records using R 3.6.0, Stats Package (R Development Core Team, 2019). Subsequently a series of characteristic curves is constructed for each test, including the tracer breakthrough curve, the corresponding tracer load curve and residence time distribution, and the tracer mass recovery curve. The tracer breakthrough curve is influenced by the amount of tracer recovered as well as by dilution effects arising from discharge variations throughout the observation period. The tracer load curve displays tracer mass flux as a function of time. It is influenced by the amount of tracer recovered, but dilution effects are eliminated, thus allowing for an unambiguous identification of multiple peaks. Integration of the tracer load curve yields the tracer mass recovery curve by successively extending the domain of integration. Normalization of the tracer load curve with respect to total recovered mass yields the residence time distribution which is neither affected by the amount of tracer recovered nor by dilution effects arising from variable discharge. Since the curve is always positive, integrable and normalized (the area under the curve equals unity) it mathematically represents a probability density distribution. Physically, it gives the residence time distribution of traced particles within the traced part of the system, i. e. integrating the residence time distribution from t_1 to t_2 gives the probability that a traced particle exhibits a residence time between t_1 and t_2. For an ideal tracer and complete recovery it gives the residence time distribution of water particles within the studies part of the system. Note, however, that with decreasing recovery rate the residence time distribution becomes less representative for the observed system.

Time to peak t_p [s] and tracer peak concentration c_p [mg m^{-3}] are directly derived from the tracer breakthrough curve. The mathematical properties of the residence time distribution allow for an analysis of its central moments (e. g. Hedderich and Sachs, 2018). The first four central moments correspond to the mean residence time $\langle t \rangle$ [s] of traced particles (first moment), the variance s^2 of residence time [s^2] (second moment), the skewness coefficient sk [−] (normalized third moment) and the kurtosis ks [−] (normalized fourth moment) of the residence time distribution, respectively (Kreft and Zuber, 1978). In case of positively skewed breakthrough curves $\langle t \rangle$ exceeds t_p, reflecting the influence of longitudinal dispersion, chemical sorption, mixing processes along the flow path, or temporary storage in stagnant water regions or the unsaturated zone (Field and Pinsky, 2000; Hauns et al., 2001; Birk, 2002; Birk et al., 2004). However, the difference should be small in case of dominant equilibrium transport in accordance with Fick's law (Field, 2002). The variance of residence time measures the spread of travel times within the aquifer. Skewness coefficient and kurtosis are dimensionless numbers quantifying the respective properties of the residence time distribution. Positive skewness indicates the distribution is weighted to the right, i. e. it recedes more gently than it rises, reflecting both longitudinal dispersion as well as potential nonequilibrium transport (Mull et al., 1988; Field, 2002). The Advection Dispersion Model (ADM) is based on the assumption of a Fickian transport process and predicts the movement and continuous spreading of a spatially symmetrical tracer cloud. Recording the passage of such a cloud yields a positively skewed breakthrough curve, since the cloud continues to spread during the passage. Note that different definitions of skewness coefficient and kurtosis are used in the literature (cf. Joanes and Gill, 1998).

The peak velocity v_p [m s^{-1}] and the mean linear velocity $\langle v \rangle$ [m s^{-1}] are derived based on the linear distance Δl, t_p and $\langle t \rangle$, respectively. Since the actual flow path length exceeds the linear distance, these velocities pose lower bound estimates on the actual flow velocities within the rock glacier. The variance of flow velocities is reflected in its dispersion characteristics. The longitudinal dispersion coefficient D [m^{2}s^{-1}] is estimated using the method of moments (Kreft and Zuber, 1978) as well as the Chatwin method (Chatwin, 1971) and results are compared. For impulse releases of tracer mass, a first-order estimate of the longitudinal dispersion coefficient is obtained employing the method of moments (Kreft and Zuber, 1978)

$$D = \frac{s^2 \langle v \rangle}{2\Delta l} \qquad \text{(Eq. 3.16)}$$

Eq. 3.16 assumes that transport is adequately described by the ADM, therefore overestimating D if the breakthrough curve is characterized by significant tailing (Field, 2002). As demonstrated by Chatwin (1971), this restriction might be circumvented by rearranging the time solution of the ADM for and impulse tracer release

$$\sqrt{t \ln \frac{\psi_2}{c\sqrt{t}}} = \frac{\Delta l}{2\sqrt{D}} - \frac{vt}{2\sqrt{D}} \qquad \text{(Eq. 3.17)}$$

(volume averaged tracer concentration c [mg m^{-3}]). The proportionality constant ψ_2 [mg s$^{0.5}$ m^{-3}] depends on injected tracer mass, the (unknown) dispersion coefficient, and the bulk flow region cross-sectional area. As demonstrated by Chatwin (1971) and Davis et al. (2000), ψ_2 might be reasonably approximated by $\psi_2 \approx c_p \sqrt{t_p}$. Strictly speaking, this approximation holds for symmetrical concentration distributions only. However, it might be used as a reasonable approximation for asymmetrical concentration distributions as well (Day, 1975; Field, 2002). Plotting the left hand side of eq. 3.17 ('Chatwin parameter') against t for each record results in a straight line of slope $-v/(2\sqrt{D})$ and intercept $(x/2\sqrt{D})$, thus allowing for an estimation of D. Significant deviations from a straight line pattern indicate that the ADM cannot represent the observations appropriately, indicating nonequilibrium processes affecting the tracer transport.

In this case, a reasonable estimate of D can still be obtained if the early data display a straight line pattern by fitting eq. 3.17 to this part only, before deviations become large and dispersion becomes increasingly dominated by non-Fickian processes (Taylor, 1954; Chatwin, 1971; Field, 2002). From these estimates for D, the corresponding dispersivities a [m] are obtained using $D = a\,v$. The dimensionless Peclet number $Pe = v\,\Delta l/D$ measures the relative influence of advective and dispersive transport, where $Pe > 6$ indicates a dominance of advective transport (Field and Nash, 1997; Field, 2002).

In addition, transport parameters are estimated by inverse modelling. A theoretical solution of the governing equation is fitted to the observed breakthrough curve using least squares regression. Two analytical models are applied, the ADM as well as the Two Region Physical Non Equilibrium Model (2RNE). Since they are based on different conceptual models of transport processes, comparing their results allows gaining insight into the processes actually taking place within the aquifer. Both approaches are based on the assumption of dominantly one-dimensional transport, stationary flow conditions, and constant aquifer properties along the flow path. However, while the ADM describes the tracer transport by advection and dispersion only, the 2RNE depicts the varying flow conditions in heterogeneous aquifers by assuming that some of the tracer is retarded in domains characterized by relatively stagnant water. During the rise in concentration (approaching tracer cloud), some of the tracer is transported into these immobile zones, temporarily retained and released again into the more mobile water zones after passage of the main tracer cloud. Assuming a first-order process to govern tracer exchange between the mobile and the immobile zone, instant homogenous distribution of the tracer within the immobile zone, and conservative tracer behavior (i. e. sorption and degradation of the tracer are negligible) the impact of this retardation is considered by adding a linear exchange term to the ADM (van Genuchten und Wagenet, 1989; Toride et al., 1999)

$$\theta_m \frac{\partial c_m}{\partial t} = \theta_m D_m \frac{\partial^2 c_m}{\partial l^2} - q \frac{\partial c_m}{\partial l} - \zeta\,(c_m - c_{im}) \qquad \text{(Eq. 3.18)}$$

(volume-averaged tracer concentration in mobile zone c_m [mg m^{-3}] and immobile zone c_{im} [mg m^{-3}], mobile zone dispersion coefficient D_m [m^2 s^{-1}], specific discharge q [m s^{-1}], first-order mass transfer coefficient ζ [s^{-1}], mobile zone volumetric water content θ_m [−]). An analytical solution for instantaneous tracer injection is provided by van Genuchten and Wagenet (1989) and implemented in CXTFIT 2.1 (Toride et al., 1999). Eq. 3.18 is transformed to a dimensionless form by assuming complete saturation at an effective porosity $\phi = 0.2$, and setting the characteristic length equal to the linear distance between injection and detection point, respectively. The resulting set of dimensionless parameters is estimated employing a nonlinear least-squares optimization approach based on the Levenberg–Marquardt method. The CXTFIT routine is used to convert the flux averaged concentrations recorded by the fluorometer to volume averaged concentrations by adjusting the boundary conditions appropriately (Kreft and Zuber, 1978; Parker and van Genuchten, 1984; Toride et al., 1993, 1999). The injected tracer mass is transformed into a concentration based on spring discharge to define the boundary condition, which is equivalent to assuming instantaneous dilution at the injection site (Birk et al., 2005). In order to ensure the iteration procedure does not converge to a local minimum, realistic initial parameters are obtained by setting $v = \langle v \rangle$ and evaluating D by eq. 3.17. The volumetric fraction of mobile water $\beta = \theta_m/\theta$ is estimated according to a rule of thumb outlined by Field and Pinsky (2000), i. e. $\beta \approx \langle v \rangle/v_p$. An initial estimate for ζ is obtained from an empirical relation for coarse grained aquifers provided by Maraqa (2001). In addition, several estimation trials are carried out with systematically varying initial estimates as recommended by Toride et al. (1999).

The derived linear velocities are used to calculate the hydraulic conductivity of the traced system assuming laminar flow conditions

$$K = \frac{v \phi \Delta l}{\Delta \varphi} \qquad \text{(Eq. 3.19)}$$

In order to examine the ratio between inertial and viscous forces affecting fluid flow, the Reynolds number Re $[-]$ is computed based on the linear velocities. For saturated flow through porous media, Re might be defined by analogy to flow through conduits using (Bear and Cheng, 2010)

$$Re = \frac{v \phi d \rho_w}{\eta} \qquad \text{(Eq. 3.20)}$$

(dynamic viscosity η [Pa s], water density ρ_w) where d [m] is some representative length characterizing the void space. However, since this length is both difficult to define exactly and different to measure appropriately, some representative measure of the grain size distribution constituting the aquifer is frequently used instead. The mean grain diameter is often taken as length dimension, although several alternative definitions are also in use (Bear, 1972). A more sophisticated analysis based on the hydraulic radius of the void space is developed by Bear and Bachmat (1990), who suggest replacing eq. 3.20 by

$$Re = \frac{\rho_w \upsilon \sqrt{\frac{k}{\phi T}}}{\eta}$$ (Eq. 3.21)

(permeability k [m^2], tortuosity T [$-$]) This definition is also suggested by Bear and Cheng (2010). Dynamic viscosity and tortuosity are approximated employing the relationships given by Busch et al. (1993), and Millington and Quirk (1961), respectively. Regardless of the exact definition used, Darcy's law is empirically found to be valid as long as Re does not exceed a value of $Re_{crit} = 1$ (Bear and Cheng, 2010).

Assuming that the rock glacier discharge is completely captured by the gauge and transport is dominated by advection (the validity of this assumption is checked by evaluating Pe), an upper bound estimate of the water volume constituting the mobile zone (2RNE) is obtained by (Atkinson et al., 1973)

$$V_m \leq \int_{t=0}^{\frac{\Delta l}{\upsilon_m}} Q dt$$ (Eq. 3.22)

Since actually several aquifer components aligned in parallel are supposed to contribute to the total spring flow, eq. 3.22 is at risk of grossly overestimating the actual mobile volume of water. Since the tracer presumably characterizes the fast flow component only, a rough correction is applied by multiplying the calculated volume by the average fraction of event water during the tracer test, as predicted by the two component mixing model based on EC (see above; Heigert, 2018; Wagner et al., 2019b).

3.6   Recharge Patterns

This section aims at deciphering characteristic recharge patterns affecting the OEG, complementing the results obtained by Heigert (2018), Winkler et al. (2018b) and Wagner et al. (2019, 2020b). Due to its large storage capacity the rock glacier mixes water from multiple recharge sources provided by the alpine environment. Potential recharge components include groundwater, rainfall, runoff from adjacent slopes, melting of snow or glacier ice, and melting of rock glacier ice (Krainer and Mostler, 2002). The alpine climate causes a strong seasonal trend in relative recharge component contributions (Krainer and Mostler, 2002, 2007; Winkler et al., 2018b; Wagner et al., 2020b). Solid precipitation dominates during the extended winter period, being stored in the snow cover. Rainfall and rising temperatures in late spring promote snowmelt, which is the governing component during early summer. As summer progresses, ice melt and rainfall recharge the rock glacier. When the temperatures fall below the freezing point in autumn, recharge ceases and spring flow is solely maintained by groundwater. The methods presented in this section provide tools for investigating the short-time fluctuations of discharge, EC, air temperature, and the way these patterns are interrelated.

Spectral analysis allows for identifying periodic variations in any time series record. Fitting sine waves by least-squares regression to the data and plotting the reduction in the sum of the residuals against frequency yields the least squares spectrum. This spectrum provides a measure of the contribution to overall variance in the data by different frequencies (Lomb, 1976; Scargle, 1982)

$$P_N(\omega) = \frac{1}{2s^2}\left(\frac{\left(\sum_j \left(y_j - \langle y\rangle\right)\cos\omega\left(t_j - \xi\right)\right)^2}{\sum\left(\cos^2\omega\left(t_j - \xi\right)\right)} + \frac{\left(\sum_j \left(y_j - \langle y\rangle\right)\sin\omega\left(t_j - \xi\right)\right)^2}{\sum\left(\sin^2\omega\left(t_j - \xi\right)\right)}\right)$$

(Eq. 3.23)

(normalized power P_N [−], total variance s^2 of time series Y consisting of data points y_j observed at times t_j, time series mean $\langle y\rangle$ ($j = 1,2,...$), angular frequency ω [s^{-1}]) where ξ is defined by $\tan 2\omega\xi = \Sigma(\sin 2\omega t_j) / \Sigma(\cos 2\omega t_j)$. The Lomb-Scargle periodogram obtained by plotting P_N against ω provides an objective tool for the detection of periodic signals despite the presence of aperiodic signals (noise) in the data series (Scargle, 1982). Eq. 3.23 yields an estimate of the power spectrum, reducing to the Fourier power spectrum in the case of equally spaced records (Lomb, 1976; Scargle, 1982). However, in contrast to the Fourier power spectrum which requires the analyzed data to be obtained at uniformly spaced intervals, the calculation of the least squares spectrum is not impaired by data gaps. Therefore, this method is appropriate for analyzing data series obtained at the rock glacier spring which exhibit several data gaps. It has been successfully applied by Birk et al. (2004) to verify localized recharge into a karst system by identifying daily periodic temperature variations.

However, the presence of noise in the record implies a periodogram that itself is noisy to some extent. The statistical significance of a prominent peak observed in the periodogram is given by the power level ς_0 derived by Scargle (1982)

$$\varsigma_0 = -\ln\left(1 - (1 - p_0)^{1/N}\right) \qquad \text{(Eq. 3.24)}$$

(number of analyzed frequencies N [−], false alarm probability p_0 [−]). Any peak exceeding this level is significant at the $(1 - p_0)$ level. This procedure ensures that the claimed detection of a periodic signal is not merely caused by random fluctuations in the data record. As demonstrated by Birk et al. (2004), combined application of eq. 3.23 and eq. 3.24 to the natural tracer record obtained at a spring allows to differentiate seasons dominated by periodic diurnal variation from those that are not. Entirely random fluctuations plot as an arbitrary number of insignificant peaks in the periodogram, while dominant periodic variations are characterized by pronounced peaks at the corresponding frequency. Since EC is regarded as a conservative tracer in crystalline catchments (Winkler et al., 2016a) the procedure might be readily adapted for the EC record of the OEG. Similarly, the procedure might be employed to detect a corresponding pattern in the spring hydrograph. Therefore, Lomb-Scargle periodograms for discharge and EC are computed using R 3.6.0 (R Development Core Team, 2019) along with the Lomb package (Ruf, 2019). To facilitate interpretation, the abscissa is scaled using periods rather than (the inversely proportional) frequencies.

As evident from fig. 3.2a, the discharge record is governed by a prominent annual periodicity. A similar (but inversely correlated) annual periodicity is recognized in the EC record (Heigert, 2018; Wagner et al., 2019b). The large variance introduced by these low frequency fluctuations impedes the identification of the comparatively less pronounced high-frequency variations (i. e. short-term fluctuations). As outlined by Birk et al. (2004), this problem can be circumvented by calculating the periodogram for the short-term fluctuations only. These are computed by filtering and removing the long-term (low frequency) fluctuations, i. e. subtracting moving averages (24 h window) from the recorded discharge and EC values, respectively. The resulting time series can be analyzed in the same way as the original series.

While traveling through the aquifer, recharge pulses are altered before finally arriving at the spring. The characteristics and extent of this alteration depend on the intensity and duration of the recharge pulse as well as on the flow geometry and hydraulic properties of the aquifer. The dimensionless response time

$$\gamma = \frac{\lambda_a}{\lambda_r} \qquad \text{(Eq. 3.25)}$$

(characteristic aquifer response timescale λ_a [s], characteristic recharge timescale λ_r [s]) introduced by Covington et al. (2009) expresses the aquifer's response to a recharge event quantitatively. It represents the relative importance of aquifer and recharge properties in altering the spring response. Responses exhibiting $\gamma < 1$ are classified as recharge dominated, i. e. the spring hydrograph shape will closely resemble the input (recharge) hydrograph shape. The geometry dominated regime is characterized by $\gamma > 1$, indicating that the shape of the recharge pulse will be significantly modified while propagating through the aquifer (Covington et al., 2009). Although originally developed for karst aquifers, the concept might be adapted to rock glaciers by employing the characteristic matrix response timescale proposed by Covington et al. (2009) for aquifer domains characterized by laminar flow

$$\lambda_a = \frac{\phi L^2}{K b} \qquad \text{(Eq. 3.26)}$$

Covington et al. (2009) define a characteristic recharge timescale equal to the width of a Gaussian input function. Similarly, the characteristic timescale λ_r [s] of meltwater recharge events driven by diurnally varying processes is suggested here to equal one day (86 400 s). A discrete event duration timescale is regarded appropriate for these recharge events since they are characterized by continuous smooth changes in flow rate (cf. Covington et al., 2009).

Cross correlation analysis allows for a comparison of time series at successive lags in order to detect positions of pronounced correspondence (Davis, 2002). The strength of the relationship is expressed by the correlation coefficient r_m [−] at match position m (Davis, 2002)

$$r_m = \frac{M \sum (Y_1 Y_2) - \sum (Y_1) \sum (Y_2)}{\sqrt{\left(M \sum (Y_1^2) - \left(\sum (Y_1)\right)^2\right)\left(M \sum (Y_2^2) - \left(\sum (Y_2)\right)^2\right)}} \qquad \text{(Eq. 3.27)}$$

for two time series Y_1 and Y_2 exhibiting M overlapping positions. Note that the summations extend only over the overlapping segments of Y_1 and Y_2. The significance of r_m may be evaluated by the approximate test (Davis, 2002)

$$t^* \approx r_m \sqrt{\frac{M-2}{1-r_m^2}} \qquad\qquad \text{(Eq. 3.28)}$$

(test statistic t^* [$-$]) having ($M - 2$) degrees of freedom. Eq. 3.27 requires a constant spacing of records throughout the observation period, thus only continuous segments of the available time series are taken into consideration (i. e. the segments between the data gaps).

Results and Interpretation

4

4.1 Internal Structure

The geometrical model is based on a high-resolution (1 x 1 m) digital elevation model of the rock glacier surface, interpolation involving 113 control points derived from the geophysical profiles, and 5510 surface displacement vectors distributed across the rock glacier surface (fig. 2.3b,e; Hausmann et al., 2012; Groh and Blöthe, 2019). Long-term creep rates are calculated using $n = 3$, $A = 2.4 \cdot 10^{-24}$ Pa^{-3} s^{-1}, $\rho_{AL} = 1604$ kg m^{-3}, and $\rho_{PF} = 1826.67$ kg m^{-3} (Konrad et al., 1999; Hausmann et al., 2012; Hartl et al., 2016). The obtained model consists of a set of four continuous surfaces which are spatially constrained by the rock glacier boundaries outlined by Wagner et al. (2019a) (fig. 2.3b). These surfaces comprise the rock glacier surface, permafrost top, permafrost base, and bedrock. They represent the interfaces of the three major hydrostratigraphic units of the OEG (active layer, permafrost layer, and base layer). The resulting elevation models are depicted in fig. 4.1 and 4.2. The projected reference coordinate system is EPSG:31254 (MGI Austria GK West), corresponding to the provided digital surface elevation model. Thus, the geometrical model is free from coordinate system transformation errors. In addition, a resampled copy at 10 x 10 m resolution is provided to facilitate integration in a potential future groundwater flow model. The downsampling procedure is carried out using ArcGIS 10.4 (cubic convolution resampling). The resampled copy aims at ensuring maximal comparability to the numerical groundwater flow model developed by Pauritsch et al. (2017) for the relict Schöneben rock glacier (Seckau Alps, Austria)[1] who used uniform cell sizes of 10 x 10 m.

---

[1] Exact location given in fig. 5.10 and tab. 5.2

© The Author(s), under exclusive license to Springer Fachmedien Wiesbaden GmbH, part of Springer Nature 2022
S. Seelig, *Characterizing Groundwater Flow Dynamics and Storage Capacity in an Active Rock Glacier*, BestMasters,
https://doi.org/10.1007/978-3-658-37073-2_4

Fig. 4.1 depicts altitude and slope angle of the calculated surfaces along with the thickness of each hydrostratigraphic unit. The **active layer** is bounded vertically by the rock glacier surface and the permafrost top. The calculated thickness ranges from 2.3 to 8.0 m and shows strong spatial heterogeneity, in good agreement with field observations at the OEG and other active rock glaciers. The mean thickness of the calculated active layer geometry is 5.3 m, slightly above the mean thickness observed along the geophysical profiles (5.2 m). Specifically, the employed calculation procedure successfully reproduces the expected correlation to the surface topography, i. e. increased thickness below ridges and edges but reduced thickness along furrows and depressions (fig. 4.1, 4.2).

The calculated **permafrost top morphology** is characterized by a set of linear features and enclosed depressions (fig. 4.1, 4.2). The longitudinal trenches are in good agreement with field observations of water running along meltwater channels eroded into the frozen rock glacier core which are traceable (visible, audible) over distances of some tens of meters along furrows in the upper part of the rock glacier (Berger et al., 2004). A set of such preferential flow paths along the modeled permafrost table is calculated using ArcGIS 10.4 and displayed in fig. 4.3. The preferential flow paths form a dendritic network that transfers runoff either to enclosed depressions or to the rock glacier front. The total volume of permafrost table depressions is evaluated by reproducing the surface with filled depressions, calculating the volume change between both surfaces and integrating the resulting layer using ArcGIS 10.4 (fig. 4.3). In summary, the depressions comprise a volume of $7.9 \cdot 10^4$ m^3, exceeding the total depression volume of $3.3 \cdot 10^4$ m^3 derived via spring flow threshold analysis by 139% (Wagner et al., 2020b). Note that the fill-spill-drain hypothesis outlined by Wagner et al. (2020b) implies that this excess is expected for two reasons: First, these depressions might already be filled to some extent prior to a precipitation event. In such a situation the actual volume that needs to be filled to induce a spring response is accordingly reduced. Second, most of the modeled depressions are located in the uppermost (southern) part of the rooting zone, while most of the allogenic recharge is expected to reach the rock glacier in the northern (lower) part of the rooting zone (cf. fig. 5.2). Thus, most of the water reaching the rock glacier will actually pass only a fraction of the permafrost table depressions displayed in fig. 4.3. Since the downsampling to 10 x 10 m smooths irregularities, the depression volume reduces to $2.5 \cdot 10^4$ m^3 for the resampled (10 x 10 m) permafrost top.

The **permafrost layer** is bounded by the permafrost top and base, respectively. Note that the modeled permafrost layer does not cover the rock glacier extent continuously throughout the rock glacier (fig. 4.1, 4.2). Its absence along

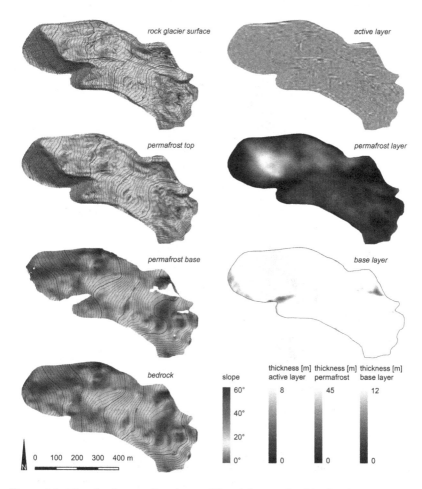

Figure 4.1   Elevation (contour lines [m a. s. l.]) and slope angle of the four interfaces (rock glacier surface, permafrost top, permafrost base, bedrock) separating the three major hydrostratigraphic units of the rock glacier (active layer, permafrost layer, base layer) derived from the geometrical model. The thickness of each hydrostratigraphic unit is also displayed (note differently scaled color codes for each unit). Note the strong heterogeneity of the active layer following the surface topography (thick below ridges, thin below furrows). Note also the pronounced thinning of the permafrost layer within the steep section associated with the bedrock swell (cf. fig. 3.1a, 4.2)

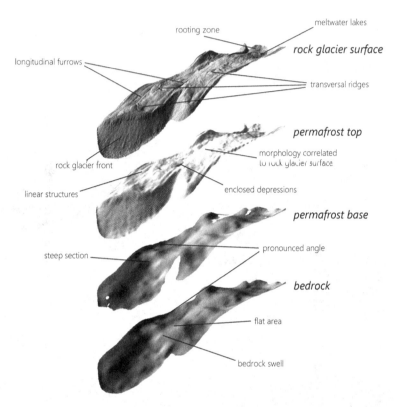

Figure 4.2   Visualization of the four interfaces separating the three major hydrostratigraphic units of the rock glacier. Elevation and slope angle of each interface are depicted in fig. 4.1

the rock glacier front and margins reflects the fact that outcropping permafrost ice is not observed at the OEG, and corresponds to the geophysical profiles depicted in fig. 3.1. The calculated permafrost base is able to satisfactorily reproduce the transition from a central flat section to the steep section below the bedrock swell observed in the geophysical profiles (compare fig. 3.1a to fig. 4.1 and 4.2). The pronounced angle along the northern margin of the flat section is also captured (compare fig. 3.1b to fig. 4.1 and 4.2). The calculated permafrost thickness pattern is in good agreement with the overall pattern of the measured Bouguer anomaly (compare fig. 2.3d to fig. 4.1). Specifically, the prominent permafrost thinning above the steep section is clearly observable (cf. fig. 3.1a). However, compared

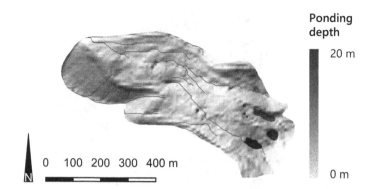

Ponding
depth

20 m

0   100 200 300 400 m

0 m

Figure 4.3 Permafrost top hydrology. Linear features promote channelized runoff. Enclosed depressions provide storage capacities in areas characterized by continuous permafrost and swallets where permafrost gets discontinuous. The indicated ponding depth corresponds to the maximal water level that might be stored within these depressions. Physical features of the modeled permafrost top geometry are depicted in fig. 4.1 and 4.2

to the gravimetry results the geometrical model seems to overestimate the permafrost thickness close to the rock glacier front (below the steep section; compare fig. 2.3d to fig. 4.1). Since the model is based on interpolated GPR and refraction seismic profiles in this part of the rock glacier, this might be either attributed to an artifact arising from the interpolation procedure, or to degrading permafrost resulting in a lower bulk density and, consequently, a less pronounced Bouguer anomaly in this part of the rock glacier.

Given the uncertainties associated with the employed creep model, the calculated permafrost thickness in the rooting zone must be regarded as a very crude approximation at best. However, the physical processes and parameters underlying the model are clearly defined, allowing for future updates of the obtained estimates using more sophisticated models (e. g. Jansen and Hergarten, 2006; Springman et al., 2012) or integrating new data if available. Note that the creep model results are incorporated in the geometrical model in the rooting zone only (due to the absence of geophysical data there, fig. 4.4). This part of the rock glacier is characterized by favorable permafrost conditions (fig. 2.3c), implying that the assumptions of the underlying creep model are more likely to be met (i. e. continuous permafrost layer, minor impact of processes such as meltwater infiltration). Anyway, regarding a prospective groundwater flow model the absolute thickness of the permafrost layer is likely to play a subordinate role due to its anticipated low specific storage. Consequently, the morphology of permafrost

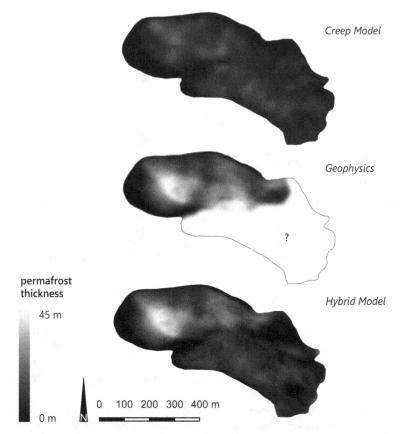

Figure 4.4   Modeled Permafrost thickness (a) based on the creep model, (b) based on inter-polation of refraction seismic and GPR profiles (c) combination of both methods, comple-menting the geophysics-based model by the creep model results where neither refraction seismics nor GPR results are available

top and base are likely to show a way higher leverage on model results than the absolute permafrost thickness.

Bedrock and permafrost base constrain the **base layer**. Since it is approximated as extending constantly 12 m below the permafrost base, its thickness is only reduced where the permafrost layer is absent (i. e. where the active layer intersects the (extrapolated) permafrost base surface defining the top of the base layer). Despite this rather artificial modification the calculated thickness simply reflects the specified conditions. The scarcity of available data implies that the obtained geometry is comparatively less certain.

4.2 Recession Analysis

The active layer is typically frozen during the winter months. Exceptions might occur on discontinuous permafrost, when the seasonal frost fails to reach the permafrost top (cf. Zhou et al., 2015). In general, however, the winter recession plausibly reflects the hydraulic characteristics of the unfrozen base layer only, since the remaining aquifer components are inactive.

Tab. 4.1 shows hydraulic conductivity estimates based on recession analysis results provided by Wagner et al. (2020b). These authors obtain a range of reasonable parameter values by combining single event analysis with analysis of the master recession curve. Accordingly, a range of hydraulic conductivity estimates is derived. In all cases, only long-term recession behavior is included in the analysis, thus the results characterize the base layer and represent lower bound estimates (as outlined above).

Table 4.1 Hydraulic Conductivity Estimates K [m s^{-1}] of the base layer, derived from the range of recession analysis results reported by Wagner et al. (2020b). K values represent lower bound estimates (volume of the associated water volume stored above spring level (minimum estimate V_{min} [m^3], maximum estimate V_{max} [m^3]; recession coefficient based on master recession curve analysis α_{MRC} [s^{-1}], based on single event analysis α_{SEA} [s^{-1}])

	α_{MRC} ($5.79 \cdot 10^{-8}$ s^{-1})	α_{SEA} ($4.63 \cdot 10^{-8}$ s^{-1})
V_{min} ($2.24 \cdot 10^5$ m^3)	$K = 2.79 \cdot 10^{-5}$ m s^{-1}	$K = 2.23 \cdot 10^{-5}$ m s^{-1}
V_{max} ($5.34 \cdot 10^5$ m^3)	$K = 2.79 \cdot 10^{-5}$ m s^{-1}	$K = 2.23 \cdot 10^{-5}$ m s^{-1}

The results given in tab. 4.1 fall into a relatively narrow range ($2.23 \cdot 10^{-5}$ – $2.79 \cdot 10^{-5}$ m s^{-1}). The geometric mean equals $2.5 \cdot 10^{-5}$ m s^{-1}. Differences arising from uncertainties of the stored volume of water (ranging from V_{min} to V_{max}) affect the 8th decimal place at the earliest (not shown). Notably, these estimates are in good agreement with hydraulic conductivities on the order of ~10^{-5} m s^{-1} characterizing the fine-grained base layer of the relict Schöneben rock glacier (Seckau Alps, Austria)[2], as determined by Pauritsch et al. (2015, 2017) and Winkler et al. (2016a). A detailed analysis of the strikingly similar

---

[2] Exact location given in fig. 5.10 and tab. 5.2

base flow behavior of both rock glaciers is given by Wagner et al. (2020b) and
Wagner et al. (in prep.).

4.3   Thermokarst Lake Water Level Fluctuations

The locally varying ice content of the ice-debris mixture constituting the rock
glacier is assumed to govern its hydraulic properties. The hydraulic conductivity of ice-saturated or ice-supersaturated parts of the rock glacier is very low,
because in this case the ice effectively seals the pores. It increases with decreasing ice content as additional percolation pathways become available. A rough
estimate is obtained from water level fluctuations of the thermokarst lake shown
in fig. 3.11 (its exact location in the rock glacier rooting zone is indicated in
fig. 2.3a). Results from the six evaluated dry summer periods free from rainfall
and snowmelt (fig. 3.12, 3.13) are summarized in tab. 4.2. The observed spread is
depicted as Box-Whisker plot in fig. 4.5. Strictly speaking, these represent upper
bound estimates, as the (unknown) lake evaporation rate is not included in the
analysis.

The resulting values range from $1.87 \cdot 10^{-9}$ m s^{-1} to $1.31 \cdot 10^{-7}$ m s^{-1}, spreading over two orders of magnitude around a geometric mean of $3.88 \cdot 10^{-8}$ m s^{-1}.
These results actually fall approximately in the upper range of hydraulic conductivity values on the order of 10^{-8} to 10^{-11} m s^{-1} reported from laboratory
permeameter tests of frozen soils (Burt and Williams, 1976; Perfect and Williams,
1980; Andersland et al., 1996; Stähli et al., 1996). The small scale of laboratory
tests compared to the 'infiltration test' adapted here (several cm^2 compared to several m^2 covered by the lake) might explain the high values (e. g. Schulze-Makuch
et al., 1999). Note that the scale of investigation of a prospective groundwater
flow model exceeds the scale of the thermokarst lake by several orders of magnitude. In addition, domains that are undersaturated with respect to ice (e. g.
as a result of degrading permafrost) are expected to exhibit strongly increased
hydraulic conductivity (Stähli et al., 1996, 1999; Zhou et al., 2015; Mohammed
et al., 2018). Therefore, the values given in tab. 4.2 should probably be regarded
as *lower* bound estimates (!) if extrapolated to a larger scale. This is particularly true for the lower part of the rock glacier where degrading permafrost is
suspected.

Table 4.2 Hydraulic Conductivity Estimates permafrost layer

Evaluation Period	Duration [hh:mm]	Water Level Difference [cm]	Hydraulic conductivity [m s^{-1}]
05.08.2015 12:00 – 10.08.2015 00:00	108:00	2.4	$6.17 \cdot 10^{-8}$
10.08.2015 12:00 – 14.08.2015 22:30	106:30	2.7	$7.04 \cdot 10^{-8}$
22.08.2015 10:30 – 23.08.2015 22:00	35:30	1.2	$9.39 \cdot 10^{-8}$
26.08.2015 10:30 – 01.09.2015 15:00	148:30	0.1	$1.87 \cdot 10^{-9}$
23.08.2016 10:00 – 28.08.2016 13:00	123:00	1.5	$3.39 \cdot 10^{-8}$
01.09.2016 15:00 – 04.09.2016 15:00	72:00	3.4	$1.31 \cdot 10^{-7}$

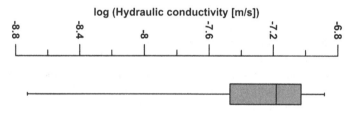

Figure 4.5 Hydraulic conductivity estimates of the permanently frozen rock glacier core beneath the thermokarst lake (sample size $= 6$)

4.4 Artificial Tracers

During both artificial tracer test (2015 and 2017) the main uranine tracer cloud passed the fluorometer within several hours, yielding tracer load curves characterized by a distinct main peak (fig. 4.6; note differently scaled axes). However, in both cases less than half of the injected uranine mass was recovered, despite the fact that the tracer concentration had nearly returned to its background level at the time the fluorometer was uninstalled (fig. 4.6). In contrast, the second tracer injected during the 2015 test (Sulforhodamine B) was not detected at the fluorometer within the observation period (145 h 27 min).

A set of characteristic curves (breakthrough curve, load curve, residence time distribution) is constructed and evaluated using the method of moments, the Chatwin method as well as inverse modelling employing the ADM and the 2RNE. The results are summarized in tab. 4.3. During the inverse modelling procedure, parameter estimation is carried out two times, setting the injected mass equal to the recovered mass and estimating the recovered mass based on the model fit. Since both approaches give very similar results, the latter is not reported.

Both tracer load curves depicted in fig. 4.6 exhibit pronounced asymmetry (tailing), suggesting that the conventional advection dispersion transport model (ADM) is not able to fully represent the flow processes within the rock glacier (note relatively high skewness coefficients in tab. 4.3). This is substantiated by the Chatwin Plots depicted in fig. 4.7. A straight line relationship is observed for early data points only. At later times (roughly exceeding the time to peak, t_p) the observed BTC clearly deviates from the shape predicted by the ADM. The observed decline in concentration during the falling limb appears to be retarded. In contrast to the ADM, the 2RNE is able to reproduce the observed BTC in good approximation (fig. 4.8; coefficient of determination $R^2_{2RNE} = 0.99$ (for both tracer tests) compared to $R^2_{ADM} = 0.80$ and 0.89 (for the 2015 and 2017 tracer test, respectively)).

Linear velocities range from $2.9 \cdot 10^{-2}$ to $6.1 \cdot 10^{-2}$ m s^{-1}, depending on the employed method of evaluation (tab. 4.3). The lowest velocities are obtained from the method of moments, reflecting the pronounced tailing of the tracer load curve (fig. 4.6, 4.7). This results in an overestimation of the first and second moment of the residence time distribution, respectively, leading to an underestimation of linear velocity. The 2RNE yields higher linear velocities than the ADM. This result is explained by the retarding effect of immobile zones, since the ADM lumps their influence with the fluid velocity in the mobile zone (Goltz and Roberts, 1986; Field and Pinsky, 2000). Regardless of the method employed, all velocities fall in the range reported from tracer tests affecting supra-permafrost flow at various active rock glaciers ($4.0 \cdot 10^{-3} - 9.1 \cdot 10^{-2}$ m s^{-1}, see below).

The longitudinal dispersion coefficient estimates based on the Chatwin method and the 2RNE are significantly lower than those obtained from the method of moments and the ADM, respectively (tab. 4.3). The Chatwin method is not affected by the breakthrough curve asymmetry, since it evaluates the rising limb only. The 2RNE explicitly accounts for the tailing. In contrast, the method of moments and the ADM integrate the asymmetry into the longitudinal dispersion coefficient, thereby inflating the estimate.

The inverse modelling results suggest that tracer transport is governed by advection, reflected in Peclet numbers > 6 (tab. 4.3). The method of moments yields lower estimates, which is a consequence of the high dispersion coefficients obtained by this method. Note the clear dominance of advective transport over diffusive transport indicated by the Peclet numbers based on the 2RNE. Hydraulic conductivity estimates range from $1.84{\cdot}10^{-2}$ to $3.48{\cdot}10^{-2}$ m s^{-1}, indicating that the tracer passes a highly permeable aquifer (tab. 4.3), in good agreement with earlier results (Rieder, 2017; Heigert, 2018) and observations at other rock glaciers (e. g. Tenthorey, 1992, 1993; Krainer and Mostler, 2002; Buchli et al., 2013; Winkler et al., 2016a). For a porous medium composed of unconsolidated sediments, this roughly corresponds to grain sizes of gravel or coarser. Note that the water flows in contact with ice, resulting in low water temperature and correspondingly high viscosity. Therefore, hydraulic conductivities are converted to permeabilities and reported separately in tab. 4.3, assuming a water density of ~1000 kg m^{-3}. The dynamic viscosity is based on the water temperature recorded at the gauge (2.7 – 2.8°C on average) and the empirical relationship provided by Busch et al. (1993). However, note that all Reynolds numbers calculated using eq. 3.21 exceed the critical Reynolds number Re_{crit} ~ 1, although not excessively. This indicates that nonlinear flow processes are likely to occur within the aquifer, questioning the validity of Darcy's law for describing the fast flow component. Eq. 3.20 cannot be evaluated directly, since the characteristic length scale (i. e. the mean grain diameter for practical purposes) is not known. However, rearranging eq. 3.20 offers a possibility to assess the flow regime at least approximately: In order to fall into the laminar flow regime, $Re < Re_{\text{crit}}$. Transforming eq. 3.20 to an inequation expected to hold as long as the flow falls within the laminar regime yields

$$\frac{\eta Re_{crit}}{v\phi\rho_w} \geq d \qquad\qquad \text{(Eq. 4.1)}$$

(Re_{crit} ~ 1). As long as the left hand side of eq. 4.1 exceeds the mean grain diameter of the aquifer, Darcy's law is not violated. The corresponding grain sizes listed in tab. 4.3 range from 0.15 to 0.28 mm, depending on the evaluation method for v. Accordingly, the mean grain diameter is not allowed to exceed the fraction of fine sand if Darcy's law is expected to hold. Taking the presence of blocks up to several meters into account, which dominate especially the uppermost part of the rock glacier (that the fast flow component is expected to pass), eq. 4.1 is most likely violated (note that huge amounts of silt or clay would be necessary to balance the impact of the large boulders; however, this is disproved by the high hydraulic conductivity values). Thus, groundwater flow in this part of the rock

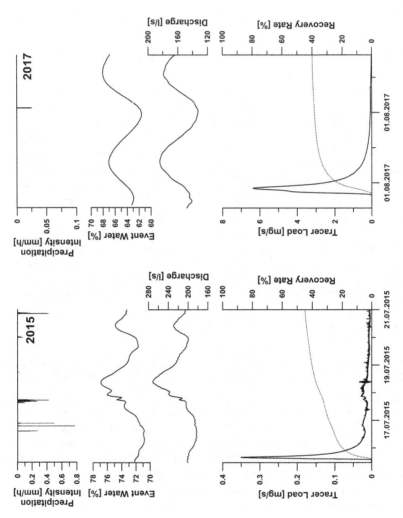

Figure 4.6 Uranine load curves 2015 and 2017. The solid line denotes the tracer load curve, while the dashed line indicates the recovery rate curve. Note differently scaled axes

glacier is likely affected by effects of inertia and turbulence, in agreement with the Reynolds numbers based on eq. 3.21.

An upper bound estimate on the volume of water constituting the mobile zone during the tracer test yields 2565 and 1780 m^3 for the 2015 and 2017 tracer test, respectively. Distributing this volume evenly across the rock glacier, assuming a drainable porosity of 0.2, results in a saturated thickness of 5.4 and 3.7 cm (!), respectively.

The estimated mass transfer coefficients ζ agree reasonably well with values reported by Pang and Close (1999) for a comparable aquifer. In addition, the results fall within the range reported from karst aquifers (1.9·10^{-6} – 4.11·10^{-5} s^{-1}; Field und Pinsky, 2000; Birk et al., 2005; Geyer et al., 2007). The obtained partition coefficient between mobile and immobile zone suggest a volumetric fraction of the mobile zone $\beta = \theta_m/\theta$ of 48% and 71%, respectively. These values might indicate a scatter around a mean of ~60%, thereby agreeing with the event water share of ~60% obtained from the two component mixing model (Heigert, 2018; Wagner et al., 2019b) as well as from the threshold analysis (Brodacz, 2019; Wagner et al., 2020b). However, this is a highly uncertain result given the small sample size and the low recovery rate.

Despite the comparable linear distance, the prominent uranine peak observed during the multi tracer test in 2015 strongly contrasts with the absence of any sulforhodamine B in the record. Possible explanations include

(1) the sulforhodamine B tracer cloud leaving the rock glacier without passing the fluorometer at the gauging station,
(2) strong dilution of sulforhodamine B resulting in concentrations below the detection limit while passing the fluorometer, or
(3) heavy retardation of the sulforhodamine B tracer e. g. by sorption processes.

Note that the creeks emerging from the five springs confluence upstream of the gauge and that slightly downstream of the gauge the creek bed is formed by outcropping bedrock (fig. 2.3a). This suggests that a large proportion of the rock glacier discharge actually passes the gauging station, thus (1) seems unlikely. As pointed out above, the detection limit of uranine is about one magnitude lower than that of sulforhodamine B. However, since the uranine peak exceeded the detection limit by nearly two orders of magnitude option (2) is regarded unlikely as well. Since the uranine was already observed after 4 h 28 min, roughly a retardation factor > 30 is needed in order to prevent the sulforhodamine cloud from reaching the fluorometer within the observation period (3). Field tracer

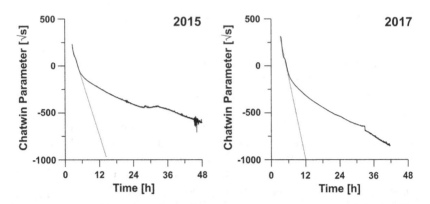

Figure 4.7 Chatwin plots of recorded uranine concentration in 2015 and 2017. Deviations from a straight-line pattern indicate non-equilibrium transport processes that are not captured by an ADM. Dotted lines indicate fits of the early-time data only, providing the basis for estimates of the longitudinal dispersion coefficient. Note strong deviations of the late-time data caused by the pronounced tailing of both breakthrough curves (cf. fig. 4.6, 4.8). Note also the slight perturbation of the early-time data, indicating a superposition of two peaks

experiments indicate retardation coefficients of sulforhodamine B with respect to uranine up to 2 (Käss, 2004). Thus, equilibrium sorption is considered unlikely to explain the apparently long travel time of sulforhodamine B with respect to uranine.

As a consequence, the most likely explanation for the high travel times of sulforhodamine B is different flow paths within the rock glacier. These might be located within the highly heterogeneous active layer or permafrost layer, with uranine travelling along preferential flow paths (e. g. channels or taliks eroded into the ice). Alternatively, the sulforhodamine B might early have reached the fine grained base layer, which exhibits a hydraulic conductivity about 3 order of magnitude lower than the active layer (see above). Similar results were obtained by Tenthorey (1992, 1993, 1994) and interpreted in this way. This interpretation is in agreement with slightly higher EC of the spring draining the southern lobe of the rock glacier (spring 1 in fig. 3.5, location indicated in fig. 2.2b), suggesting that the rock glacier consists of distinct hydraulic domains (Heigert, 2018).

The low uranine recovery rate of both tracer tests might be caused by several factors (or any combination there of):

Table 4.3 Tracer test results (dispersivity a [m^2] estimate based on advection dispersion model (a_{ADM}), based on Chatwin method (using the intercept $a_{Ch,int}$, using the slope $a_{Ch,slp}$, average of both estimates $\langle a_{Ch} \rangle$), based on method of moments (a_{MM}), based on two region physical nonequilibrium model (dispersivity of mobile zone a_m); tracer peak concentration c_p [mg m^{-3}]; longitudinal dispersion coefficient D [m^2 s^{-1}] based on advection dispersion model (D_{ADM}), based on Chatwin method (using the intercept $D_{Ch,int}$, using the slope $D_{Ch,slp}$, average of both estimates $\langle D_{Ch} \rangle$), based on method of moments (D_{MM}), based on two region physical nonequilibrium model (longitudinal dispersion coefficient of mobile zone D_m); critical mean grain diameter for laminar flow $\langle d \rangle_{lam}$ [mm] based on advection dispersion model ($\langle d \rangle_{lam,ADM}$), based on method of moments ($\langle d \rangle_{lam,MM}$), based on two region physical nonequilibrium model ($\langle d \rangle_{lam,2RNE}$); hydraulic conductivity K [m s^{-1}] estimate based on advection dispersion model (K_{ADM}), based on method of moments (K_{MM}), based on two region physical nonequilibrium model (K_{2RNE}); permeability k [m^2] estimate based on advection dispersion model (k_{ADM}), based on method of moments (k_{MM}), based on two region physical nonequilibrium model (k_{2RNE}); kurtosis of residence time distribution kt [-]; injected tracer mass m_{in} [g]; recovered tracer mass m_{rec} [g]; recovery rate $m_\%$ [%]; Peclet number Pe [-] estimate based on advection dispersion model (Pe_{ADM}), based on method of moments (Pe_{MM}), based on two region physical nonequilibrium model (Pe_{2RNE}); mean spring discharge $\langle Q \rangle$ [l s^{-1}]; coefficient of determination R^2 [-]; Reynolds number Re [-] estimate based on advection dispersion model (Re_{ADM}), based on method of moments (Re_{MM}), based on two region physical nonequilibrium model (Re_{2RNE}); standard deviation of residence time distribution sd [hh:mm]; skewness coefficient of residence time distribution sk [-]; time to peak t_p [s]; mean residence time $\langle t \rangle$ [s]; volume of mobile water V_m [m^3]; linear velocity v [m s^{-1}] estimate based on advection dispersion model (v_{ADM}), based on method of moments ($\langle v \rangle$), based on two region physical nonequilibrium model (v_m); peak velocity v_p [m s^{-1}]; linear distance x_{lin} [m]; vertical distance x_v [m]; partition coefficient for mobile and immobile zones β [−]; mass transfer coefficient ζ [s^{-1}]; water temperature $\langle \vartheta \rangle$ [°C]).

Parameter	Value 2015	Value 2017
Tracer Injection	15.07.2015 11:20	31.07.2017 17:08
Duration [hh:mm]	145:33	41:44
m_{in} [g]	25	202.49
x_{lin} [m]	980	910
x_v [m]	313	284
$\langle Q \rangle$ [l s^{-1}]	191	160
$\langle \vartheta \rangle$ [°C]	2.7	2.8
m_{rec} [g]	11.46	80.90
$m_\%$ [%]	45.83	39.95
t_p [hh:mm]	04:28	05:20
c_p [mg m^{-3}]	1.75	42.32
v_p [m s^{-1}]	0.061	0.047
V_m [m^3]	2565	1780

(continued)

Table 4.3   (continued)

Parameter	Value 2015	Value 2017
Method of Moments		
$\langle t \rangle$ [hh:mm]	09:16	08:41
sd [hh:mm]	06:00	06:09
sk [$-$]	1.29	2.59
kt [$-$]	3.75	10.23
$\langle v \rangle$ [m s^{-1}]	0.029	0.029
D_{MM} [m^2 s^{-1}]	6.05	6.65
a_{MM} [m]	206	229
K_{MM} [m s^{-1}]	$1.84 \cdot 10^{-2}$	$1.86 \cdot 10^{-2}$
k_{MM} [m^2]	$3.05 \cdot 10^{-9}$	$3.08 \cdot 10^{-9}$
Pe_{MM} [$-$]	4.76	3.98
Re_{MM} [$-$]	2.91	2.91
$\langle d \rangle_{lam,MM}$ [mm]	0.277	0.279
Chatwin Method		
$\langle D_{Ch} \rangle$ [m^2 s^{-1}]	0.73	0.24
$D_{Ch,int}$ [m^2 s^{-1}]	1.18	0.35
$D_{Ch,slp}$ [m^2 s^{-1}]	0.28	0.13
$\langle a_{Ch} \rangle$ [m]	25	8
$a_{Ch,int}$ [m]	40	12
$a_{Ch,slp}$ [m]	10	4
Advection Dispersion Model (ADM)		
R^2	0.80	0.89
v_{ADM} [m s^{-1}]	0.035	0.042
D_{ADM} [m^2 s^{-1}]	5.47	1.37
a_{ADM} [m]	158	33
K_{ADM} [m s^{-1}]	$2.17 \cdot 10^{-2}$	$2.69 \cdot 10^{-2}$
k_{ADM} [m^2]	$3.61 \cdot 10^{-9}$	$4.46 \cdot 10^{-9}$
Pe_{ADM} [$-$]	6.22	27.85
Re_{ADM} [$-$]	3.47	5.05
$\langle d \rangle_{lam,ADM}$ [mm]	0.235	0.193
Two Region Physical Nonequilibrium Model (2RNE)		
R^2	0.99	0.99

(continued)

Table 4.3 (continued)

Parameter	Value 2015	Value 2017
v_m [m s^{-1}]	0.056	0.047
D_m [m^2 s^{-1}]	1.58	0.49
a_m [m]	24	12
β [−]	0.48	0.71
ζ [s^{-1}]	5.70·10^{-6}	4.51·10^{-6}
K_{2RNE} [m s^{-1}]	3.48·10^{-2}	3.00·10^{-2}
k_{2RNE} [m^2]	5.78·10^{-9}	4.96·10^{-9}
Pe_{2RNE} [−]	40.35	75.25
Re_{2RNE} [−]	7.59	5.93
$\langle d \rangle_{lam,2RNE}$ [mm]	0.146	0.174

(1) Degradation due to exposure to sunlight between the rock glacier spring and the gauging station (150 m distance). However, assuming a velocity of 0.3 m s^{-1}, this might cause a degradation of 10% at maximum (employing the relationship given by Leibundgut et al., 2009). Note also that during the 2017 test, the peak occurred at night (22:30), precluding degradation during that test (during the 2015 test, the peak occurred at 16:00, indicating that photolytic degradation is possible to some extent).

(2) A large proportion of the tracer does not pass the gauging station after leaving the rock glacier and is therefore not recorded. This option has already been ruled out above.

(3) A portion of the tracer mass is transported into areas where direct exchange with water moving along the fast flow paths is inhibited. This is in good agreement with the fill-spill-drain hypothesis (Wagner et al., 2020b) and closely related to the observed tailing (see below). Note also that the frontal part of the rock glacier is under tensional strain (fig. 2.3e, f) promoting the development of crevasses. As a consequence of rapid movement, this part of the rock glacier is also at or below the lower permafrost boundary as indicated by Boeckli et al. (2012a,b) (Fig. 2.3c), promoting the development of thermokarst features (e. g. taliks). These features promote temporary retardation of a part of the tracer mass or transfer it to the lowly permeable base layer, preventing it from reaching the fluorometer within the observation period. This is regarded as the most plausible explanation.

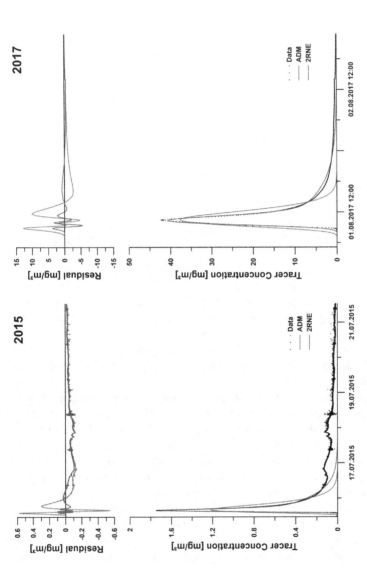

Figure 4.8 Results of inverse modelling for the 2015 tracer tests (left) as well as for the 2017 tracer test (right) employing the ADM (red) and the 2RNE (blue). The 2RNE is able to capture the tailing of both breakthrough curves appropriately, resulting in a satisfactorily coefficient of determination $R^2_{2RNE} = 0.99$ (for both tracer tests). The ADM is not able to account for the nonequilibrium transport processes taking place within the rock glacier (cf. fig. 4.7), resulting in lower coefficients of determinations $R^2_{ADM} = 0.80$ and 0.89 (for the 2015 and 2017 tracer test, respectively)

Note that both tracer load curves show indications of multiple peaks, although they are not distinct. These affect the rising limb in 2015 and 2017 (fig. 4.6), more easily recognized in the perturbation of the early straight line pattern in fig. 4.7. In 2015, two additional peaks occur at later times (note the low overall uranine concentration in 2015, introducing some uncertainty). Since dilution effects arising from varying discharge are removed in these curves, the additional peaks are attributable to features of the aquifer, indicating a highly heterogeneous structure. However, the two late-time peaks (at least the second) observed in 2015 might also be attributed to recharge pulses stemming from infiltrating precipitation water (fig. 4.6) remobilizing some temporarily retarded tracer; cf. Winkler et al., 2016; Seelig, in prep.). Similar results are obtained by Tenthorey (1992, 1993) and Pedevilla (2019).

The tailing of both tracer breakthrough curves indicates that some non-equilibrium process retarding the tracer transport is taking place within the rock glacier (fig. 4.6–4.8). As pointed out above, uranine is not heavily affected by chemical sorption. Due to the presence of a perched aquifer on top of the frozen rock glacier core, the percolation pathways of infiltrating water through the unsaturated zone are short and, consequently, excessive temporary storage in the unsaturated zone is considered unlikely. A possible process causing the tailing that is consistent with the highly heterogeneous structure of the rock glacier is the presence of stagnant water within the part of the system characterized by the tracer test. Within the active rock glacier, stagnant water may be present in caverns eroded into the ice core, behind boulders or ice protrusions, crevasses or low permeability domains (fine grained or partially frozen sediment). Specifically, the spill-fill-drain hypothesis suggests the presence of topographic depressions or pool structures at interfaces of strongly contrasting permeability as the permafrost top or bedrock (Wagner et al., 2020b). Due to thermal constraints, the top of the permafrost body is correlated to some extent to the surface morphology, promoting the occurrence of topographic depressions (fig. 4.2, 4.3). If a significant portion of the fast-flow component is flowing on top of the frozen permafrost body, these depressions might act as temporary storage zones. Conceptually this agrees with presence of features promoting water stagnation, thereby retarding flow. Note that some of the tracer migrating into these zones might be transported diffusively into low-permeability domains, thereby explaining the low recovery rate (due to strong dilution or retardation times exceeding the tracer test duration).

Alternatively, mixing processes and secondary flows caused by vortices and eddies due to surface irregularities along flow margins, abruptly changing flow channel cross-section or anastomosing channel patterns, hydraulic jumps, or dead-end passages might be responsible for the observed tailing. These processes are likely to influence the transport pattern to a certain extent if part of the traced water is moving along channels eroded into the ice core, as suggested by field observations (Berger et al., 2004), the high velocities and Reynolds numbers indicating a nonlinear flow regime (see above), and the recharge-dominated spring response (see below). Similar processes are considered to induce highly asymmetrical breakthrough curves in karst systems (c. f. Jeannin and Maréchal, 1998; Field and Pinsky, 2000; Hauns et al., 2001; Birk et al., 2005; Geyer et al., 2007). Subsequent to parameter optimization a sensitivity analysis is conducted. Results obtained from a systematic variation of individual parameters are depicted in fig. 4.9, indicating a strong sensitivity of the model with respect to v and β. The results are very similar to those obtained by Geyer et al. (2007) and therefore not repeated here. In addition, the correlation matrices given by

$$
\begin{array}{cc}
\quad v \quad D \\
v \quad 1 \\
D \; -0.310 \; 1
\end{array}
\qquad
\begin{array}{cc}
\quad v \quad D \\
v \quad 1 \\
D \; -0.179 \; 1
\end{array}
$$

for the fitted (dimensionless) ADM parameters (2015 left, 2017 right) and by

$$
\begin{array}{ccccc}
& v & D & \beta & \Omega \\
v & 1 \\
D & 0.108 & 1 \\
\beta & \mathbf{0.880} & 0.470 & 1 \\
\omega & 0.227 & -\mathbf{0.674} & -0.154 & 1
\end{array}
\qquad
\begin{array}{ccccc}
& v & D & \beta & \Omega \\
v & 1 \\
D & 0.035 & 1 \\
\beta & \mathbf{0.950} & 0.254 & 1 \\
\omega & 0.221 & -\mathbf{0.718} & -0.023 & 1
\end{array}
$$

for the dimensionless 2RNE parameters reveal a strong positive correlation between v and β as well as between D and the dimensionless mass exchange coefficient $\omega = (\zeta\, L)/(\theta\, v)$ (i. e., linearly correlated to ζ). While the parameter estimation process is able to provide reliable combination of these parameters (indicated by the good overall fit achieved by the model), individual values are less certain. A simultaneous change in strongly correlated parameters hardly affects the objective function due to their similar effect on the tracer breakthrough curve: *Higher* flow velocity and *lower* mobile water volume both shift the peak of the modeled breakthrough peak to the left, resulting in a strongly *positive* correlation between v and β. Similarly, the curve flattens due to *increased* dispersion as well as due to *increased* rate of mass transfer between mobile and immobile zone, implying a *negative* correlation between D and ω (resulting in the same correlation between D and ζ due to the linear correlation of ω and ζ). However,

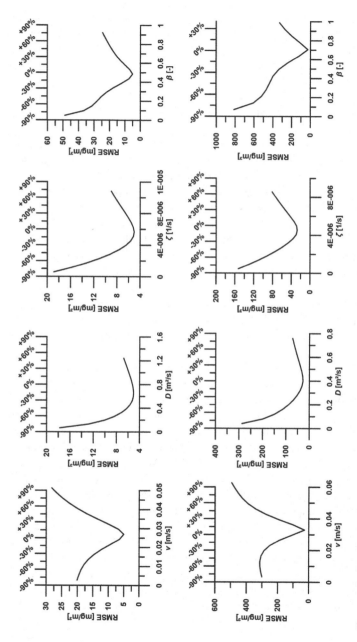

Figure 4.9 Sensitivity analysis of the 2015 (upper part) and 2017 (lower part) 2RNE model, indicating a strong sensitivity of the model with respect to v and β. Similar patterns were derived by Geyer et al. (2007)

according to a rule of thumb provided by Hill and Tiedeman (2007) parameter correlations < 0.95 indicate that parameters can be estimated uniquely in general. Inspection of the parameter correlation matrices shows that all of the observed correlation coefficients are less than or equal this threshold.

4.5   Recharge Patterns

Glaciers and rock glaciers gain ice from snow metamorphism, surface icing, and (re-)freezing of percolating water (Haeberli et al., 2006; Cuffey and Paterson, 2010). The ice surface loses mass by ablation and sublimation caused by radiation and turbulent mixing of heat and vapour in the adjacent air (Cuffey and Paterson, 2010). The melting rates exhibit strong variations in response to diurnal and seasonal cycles of radiation and temperature as well as to local weather conditions. During fair weather conditions, the periodic diurnal variation of radiation and temperature imprints a corresponding periodic variation in discharge of meltwater creeks. Since snowmelt and ice melt are the only recharge components driven by these processes, recharge during seasons dominated by meltwater should exhibit periodic variations in discharge by a wavelength of approximately one day (preserving the wavelength of the corresponding melting process). When snowfall shuts down the melting processes, the diurnal variations in discharge are expected to level off.

Fig. 4.10a and 4.10b display the periodograms of the complete discharge and EC record, respectively. Clearly, both time series are dominated by an annual periodic variation, corresponding to the prominent peak at a period of 365 days in both periodograms. Additional peaks at multiples of the corresponding fundamental frequency are attributed to spectral leakage (e. g. at ½ year; Scargle, 1982). These results confirm the annual periodicity observed in fig. 3.2a and in the corresponding EC record (cf. Heigert, 2018; Wagner et al., 2019a), which is in good agreement with previous findings (e. g. Krainer and Mostler, 2002; Krainer et al., 2007). However, the time series variance resulting from this seasonality masks the impact of potential short-term periodicity on the overall time series. This problem is solved by filtering the low frequencies as outlined above. The resulting periodograms are given in fig. 4.10c and 4.10d. Evidently, short-term fluctuations are dominated by diurnal periodic fluctuations, since a single prominent peak corresponds to a wavelength of one day. Note that this result is not restricted to selected seasons but takes the complete record into account. Consequently diurnal variations are the only periodic signals affecting the short

term fluctuation, leading to the conclusion that melting processes (suspected to
be the only processes causing periodic fluctuations with a wavelength of one day)
govern any periodic short-term variation.

These observations can be quantified by calculating the corresponding power
level ς_0 at a specified false alarm probability. In Fig. 4.10 – 4.12 the power
level at a false alarm probability $p_0 = 10^{-5}$ is plotted as dotted line, allowing these figures to be used as simple statistical tests. The results can be read
off the graphs directly and immediately quantified. The null hypothesis states
that recorded fluctuations are completely aperiodic, implying that peaks observed
in the periodogram are merely a random superposition of arbitrary frequencies.
Conversely, eq. 3.24 ensures that any peak exceeding the plotted power level
(dotted line) is statistically significant at the 0.99999 level (Scargle, 1982). Thus,
fig. 4.10c and 4.10d demonstrate that diurnal variations are the only significant
periodic short-term fluctuations influencing the filtered discharge and EC record.
Note that this procedure is applicable regardless of the exact type of signal studied (i. e. might be used for any natural or artificial tracer as long as physically
plausible).

As demonstrated by Birk et al. (2004), daily periodic variation in natural
tracers might indicate a continuing localized infiltration. In contrast to rainfall
and groundwater recharging the rock glacier, the meltwater creek infiltrating into
the rock glacier rooting zone represents such a localized infiltration. Mixing of
meltwater with rainfall or irregular surface runoff disturbs the periodic trend,
implying less pronounced power of the corresponding period (1 day) in the dia-
gram. Similarly, reduced or impeded meltwater production results in the absence
of a significant peak at this period. In turn, seasons dominated by meltwater
are identified by particularly pronounced peaks. Following Birk et al. (2004), the
least squares spectra of short-term fluctuations are calculated for each month separately. The results are plotted in fig. 4.11 (discharge) and fig. 4.12 (EC), with
power levels plotted at the 0.99999 level. If less than 50% of the month is covered by a corresponding record, the respective periodograms are light-colored.
All records show pronounced peaks at a period of one day during May, June,
July, and August, sometimes also in April. Occasionally, these peaks are found
in March, September, or October. These prominent peaks correspond to months
dominated by snow melt (early summer) and/or ice melt (late summer). Periodograms exhibiting random noise (November, December, January, February;
often March and October, sometimes April and September) indicate that these
contribute only insignificant amounts or are absent. The 2014 discharge record

deviates somewhat from this pattern, exhibiting a damped impact of diurnal fluctuations during August but pronounced diurnal variation in October. This reflects the cold and wet conditions prevailing during late summer 2014 and the following warm autumn, with temperature remaining anomalously high until ~20.10.2014.

Colombo et al. (2018a) showed that solute export from an active rock glacier is driven primarily by rainfall rather than by air temperature. In contrast to air temperature, rainfall does not exhibit a distinct periodic diurnal variation. This suggests that the periodic diurnal variations in discharge and EC are related to snow melt or glacial meltwater, suggesting that meltwater constitutes a significant recharge component due to the pronounced impact of diurnal variations on discharge and EC. The individual months characterized by strong melting processes can be identified using fig. 4.11 and fig. 4.12 (prominent peak at period of 1 day). The fact that this impact is sustained until early autumn (i. e. long after the snow cover has completely molten) suggests that ice melt accounts for much of this meltwater contribution. This is in good agreement with results of the rainfall runoff model (fig. 3.7, 5.1).

Fig. 3.6c indicates that during warm, dry summer periods, discharge is inversely correlated to EC, both showing periodic variation at a period of one day (cf. fig. 3.4). In addition, discharge is directly correlated to air temperature (AT) with a time lag of 16 h. Both correlations are expressed quantitatively using cross-correlograms (fig. 4.13). Positions of maximum correspondence are summarized in tab. 4.4 and evaluated for their significance. The discharge-EC correlation (fig. 4.13a) reaches a minimum $r_0 = -0.70$ at a lag time of 0 h. Thus, there is no temporal offset between discharge maximum and EC minimum (Note that due the sinuous shape of both time series, the inverse correlation approaches local minima (less significant) at lag times of -24 h and $+24$ h, respectively. For the same reason, local maxima (less significant) appear at lag times of -13 h and $+12$ h, respectively). This inverse correlation becomes almost perfect if only short-term fluctuations of both time series are considered. The long-term variation is filtered according to the procedure outlined above. The resulting time series can be correlated in the same way as the original time series. The obtained correlogram (fig. 4.13b) reveals an extremely strong negative correlation ($r_0 = -0.98$) at a lag time of 0 h.

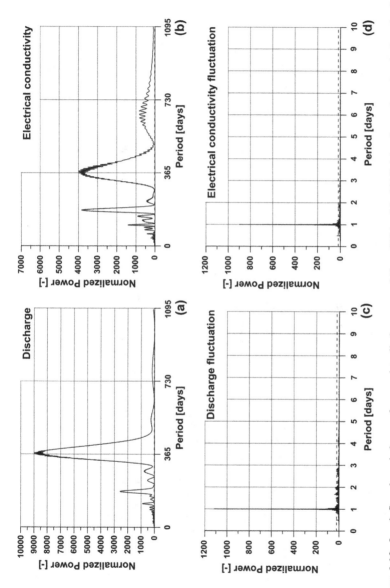

Figure 4.10   Lomb-Scargle periodogram for complete discharge record (June 2014 – July 2018) (a) and complete electrical conductivity record (July 2015 – July 2018) (b), both dominated by annual periodicity (365 days). Diurnal variations (period of 1 day) clearly dominate short-term fluctuations in discharge (c) and electrical conductivity (d). Peaks above the dotted line are significant at the 0.99999 level

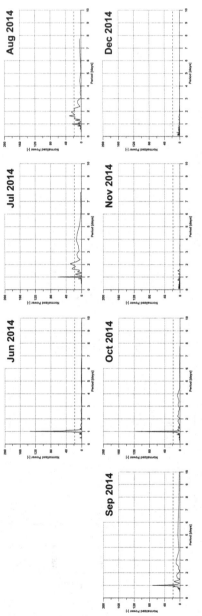

Figure 4.11 Lomb-Scargle periodograms for short-term fluctuations in discharge during 2014, computed per month. Light periodograms indicate data time series covers less than 50% of month. Peaks above the dotted line are significant at the 0.99999 level

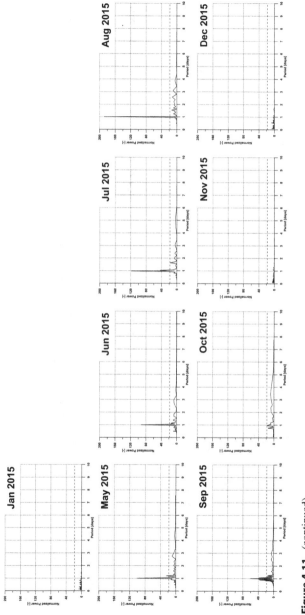

Figure 4.11 (continued)

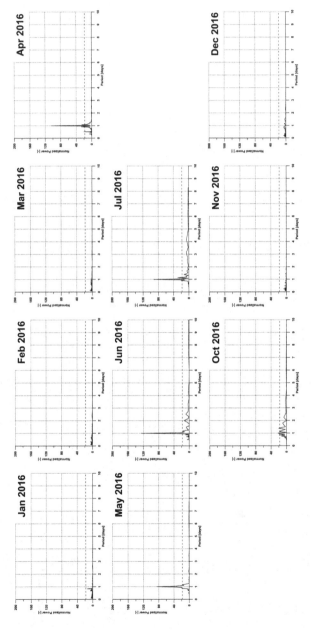

Figure 4.11 (continued)

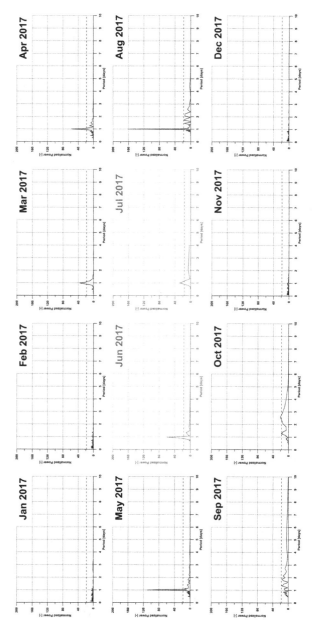

Figure 4.11   (continued)

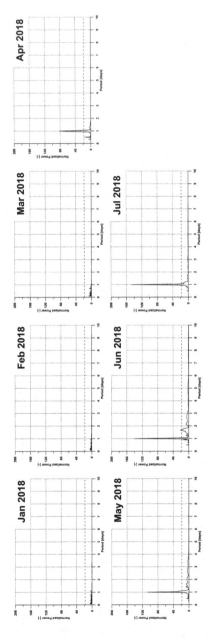

Figure 4.11 (continued)

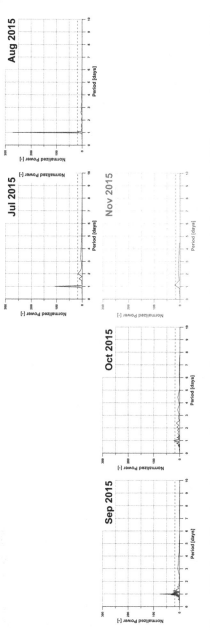

Figure 4.12 Lomb-Scargle periodograms for short-term fluctuations in electrical conductivity during 2015, computed per month. Light periodograms indicate data time series covers less than 50% of month. Peaks above the dotted line are significant at the 0.99999 level

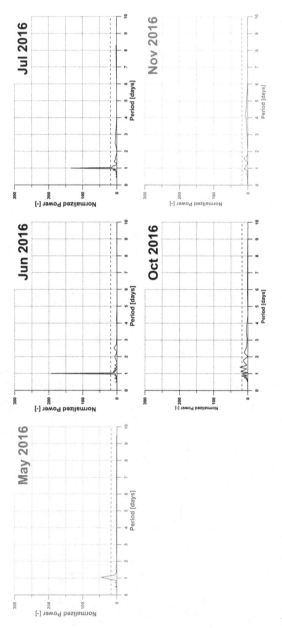

Figure 4.12 (continued)

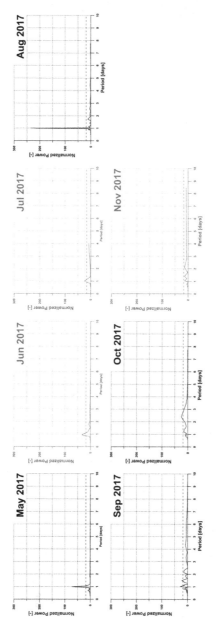

Figure 4.12 (continued)

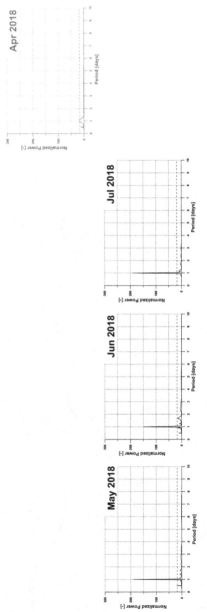

Figure 4.12 (continued)

The correlation between discharge and air temperature is somewhat less distinct (fig. 4.13c). Again, it is enhanced if only short-term fluctuations are considered (fig. 4.13d), yielding local maxima at -8 h (0.80) and $+16$ h (0.79) and local minima at -20 h (-0.75) and $+4$ h (-0.83), respectively. Some care is needed when interpreting these results, taking the effects arising from the sinusoidal shape of both records into account. Clearly, air temperature influences discharge (and not vice-versa), precluding negative lags. Assuming that the diurnal variations are caused by some melting process, negative correlations are physically not meaningful, leaving only the maximal positive correlation at $+$ 16 h ($r_{16} = 0.79$). This indicates that during warm, dry summer periods, glacial meltwater causes peaks in discharge about 16 h after the corresponding peaks in air temperature (cf. Wagner et al., 2019a). This time lag equals the travel time of meltwater from its source to the gauge. Note that the apparent inverse correlation between discharge and air temperature in fig. 3.6c is rather random (air temperature rises more rapidly than it declines, thus the daily minima follow the maxima at a lag of $>$ 12 h).

The inverse correlation between discharge and EC also holds if the complete record is considered. Fig. 4.14 and 4.15 depict the corresponding cross-correlation for six periods between summer 2015 and 2018, respectively. For these periods plausible discharge and EC time series are available (a more detailed analysis is given by Heigert, 2018, and Wagner et al., in prep.). All six periods exhibit a maximum inverse correlation at a lag of ~0 h (fig. 4.14). However, the correlation clearly gains significance if only short-term fluctuations are considered (fig. 4.15; still maximum inverse correlation at a lag of 0 h). Fig. 4.14 and 4.15 might be interpreted with more confidence than fig. 4.13, since the covered periods are long compared to the analyzed lag times (cf. Box et al., 2016).

The strong correlation between air temperature and spring discharge suggests an accordingly strong correlation between the meltwater recharge (input) hydrograph and the spring discharge. Based on the spectral analysis and cross correlation analysis results, the dimensionless response time is clearly $\gamma < 1$ although its precise value remains unknown, indicating a recharge dominated spring response (Covington et al., 2009). Accordingly, eq. 3.25 predicts $\lambda_a < \lambda_r$; yielding an upper bound constraint on the characteristic aquifer response timescale $\lambda_a <$ 1 day. However, employing the results of the artificial tracer tests (which characterize the fast flow component), eq. 3.26 predicts λ_a between 810 and 3158 days (depending on the exact values taken from the artificial tracer test results, see tab. 4.3), dramatically exceeding this upper bound constraint. This obvious contradiction indicates that the conceptual model attributing the fast flow component to a highly permeable, homogeneous porous medium is inappropriate. Instead, this aquifer component is probably broken up into smaller matrix blocks, separated by preferential flow paths exhibiting significantly lower response timescales (Covington et al., 2009). This is in perfect agreement with the observed meltwater channels eroded into the OEG permafrost table (Berger et al., 2004). Similar observations are reported from many active rock glaciers (see below; Wahrhaftig and Cox, 1959; Potter, 1969, 1972; White, 1971; Johnson, 1978; Giardino et al., 1992; Tenthorey, 1992, 1993; Krainer and Mostler, 2002; Vonder Mühll et al., 2003; Arenson et al., 2010; Springman et al., 2012; Buchli et al., 2013, 2018). This result substantiates suggestions by Winkler et al. (2016a,b); Pauritsch et al. (2017), Heigert (2018); Wagner et al. (2020b) who emphasized the complex system response of rock glaciers similar to karst systems (note that some of these publications refer to relict rock glaciers).

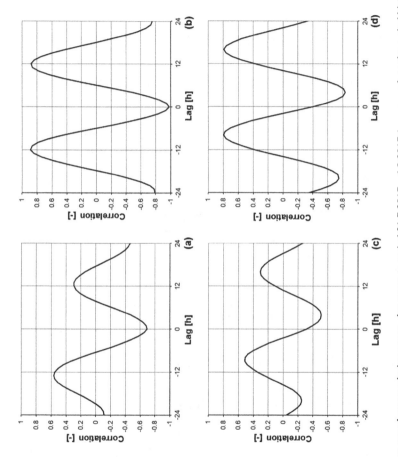

Figure 4.13 Cross-correlograms during warm, dry summer period 31.7.2017 – 6.8.2017 (corresponds to the period highlighted in fig. 3.4 and displayed in fig. 3.6c). Cross-correlation of (a) discharge and electrical conductivity, (b) fluctuations in discharge and electrical conductivity (c) discharge and air temperature (d) fluctuations in discharge and air temperature

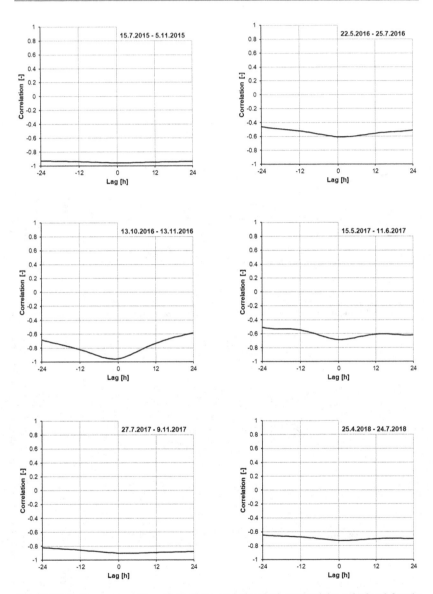

Figure 4.14 Cross correlation of discharge and electrical conductivity calculated for six periods of plausible, continuous record

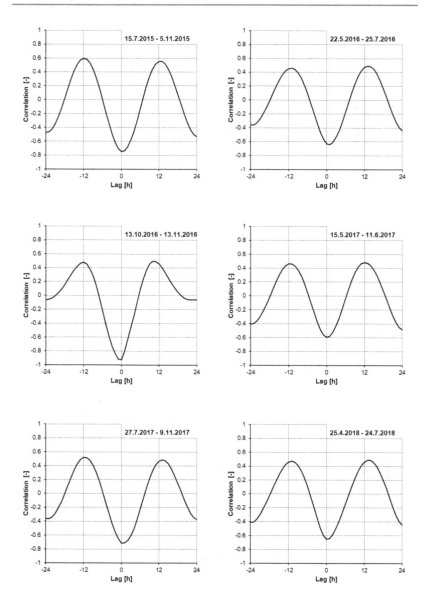

Figure 4.15   Cross-correlation between short-term fluctuations in discharge and electrical conductivity, respectively

Table 4.4 Cross correlation significance

Correlated Series	Period	Correlegram	r_m [−]	lag [h]	Length of record [days]	test statistic t^* [−]	critical value at 0.99 level	result
Q/EC	31.7.2017 – 6.8.2017	fig. 4.13a	−0.7	0	6	−12.12	−2.3510	significant
Q/EC	31.7.2017 – 6.8.2017	fig. 4.13b	−0.98	0	6	−60.91	−2.3510	significant
Q/AT	31.7.2017 – 6.8.2017	fig. 4.13c	0.31	16	6	−3.82	−2.3539	significant
Q/AT	31.7.2017 – 6.8.2017	fig. 4.13d	0.79	16	6	−15.08	−2.3539	significant
Q/EC	15.7.2015 – 5.11.2015	fig. 4.14a	−0.958	0	113	−173.59	−2.3277	significant
Q/EC	22.5.2016 – 25.7.2016	fig. 4.14b	−0.607	1	64	−29.99	−2.3288	significant
Q/EC	13.10.2016 – 13.11.2016	fig. 4.14c	−0.962	−1	31	−95.32	−2.3315	significant
Q/EC	15.5.2017 – 11.6.2017	fig. 4.14d	−0.685	0	28	−24.19	−2.3320	significant
Q/EC	27.7.2017 – 9.11.2017	fig. 4.14e	−0.902	3	105	−104.55	−2.3278	significant
Q/EC	25.4.2018 – 24.7.2018	fig. 4.14f	−0.726	1	90	−49.14	−2.3281	significant
Q/EC	15.7.2015 – 5.11.2015	fig. 4.15a	−0.747	0	113	−58.38	−2.3277	significant
Q/EC	22.5.2016 – 25.7.2016	fig. 4.15b	−0.642	1	64	−32.88	−2.3288	significant
Q/EC	13.10.2016 – 13.11.2016	fig. 4.15c	−0.926	0	31	−66.32	−2.3315	significant
Q/EC	15.5.2017 – 11.6.2017	fig. 4.15d	−0.594	0	28	−19.00	−2.3320	significant
Q/EC	27.7.2017 – 9.11.2017	fig. 4.15e	−0.712	0	105	−50.77	−2.3278	significant
Q/EC	25.4.2018 – 24.7.2018	fig. 4.15f	−0.649	0	90	−39.72	−2.3281	significant

Discussion

5

5.1 Conceptual Model

The ultimate goal in alpine catchment characterization is to predict hydrological responses at the watershed scale. First-order controls on the hydrological behavior of the Innere Ölgrube are identified by Wagner et al. (in prep.). The integration of a distributed parameter model focusing on dominant processes at the rock glacier scale into this generic framework requires a coordinated evaluation of models characterizing different scales (Grayson and Blöschl, 2000; Sivapalan et al., 2003). Therefore this thesis suggests starting from the rainfall runoff model results provided by Wagner et al. (in prep.) and building complexity carefully in an iterative procedure. The temporal distribution of rainfall, snow melt and ice melt as recharge components inferred by Wagner et al. (in prep.) is depicted in fig. 5.1 (cf. fig. 3.7). Specifying the boundary conditions of a prospective groundwater flow model of the OEG additionally requires the spatial distribution of these components.

Relative contributions of individual recharge components change on a seasonal as well as on a diurnal time scale (fig. 3.4, 3.6, 3.7, 5.1; Krainer and Mostler, 2002; Winkler et al., 2018a,b; Wagner et al., 2019a, 2020b). Snow melt and rainfall dominate during early summer, shifting to ice melt and rainfall once the snow cover has completely molten. Groundwater becomes increasingly important during late summer and early autumn, when temperatures start to decline. Recharge is negligible during the winter recession (snow cover), where spring flow is sustained by water stored in the rock glacier (Wagner et al., 2020b). Changing seasonal (and diurnal) contributions from these sources and local weather conditions affect flow paths and storage capacities: Heat transport by flowing water and circulating air governs the seasonal variation in active layer thickness, promotes water infiltration into the frozen rock glacier core and the development of thermokarst features during the summer.

Most of the catchment area above the rock glacier consists of bare rock assumed to have negligible permeability and storage capabilities. Consequently, most of the rainfall and snow melt in these domains reaches the rock glacier as surface runoff. Based on the approach taken by Pauritsch et al. (2017), the catchment is compartmentalized into several subcatchments (fig. 5.2). The total rainfall and snow melt flux within the catchment provided by Wagner et al. (2020b) (fig. 3.7, 5.1) is distributed among these subcatchments in proportion to their respective horizontal area. However, glacial meltwater reaching the rock glacier as allogenic recharge arises exclusively in subcatchment z3. In contrast, melting of permafrost ice results in distributed production of meltwater across wide parts of the catchment, including the rock glacier itself (recharge zone z1 in fig. 5.2; cf. fig. 2.3c). Thus, the ice melt contribution evaluated by Wagner et al. (2020b) needs to be analyzed more closely to infer its spatial distribution.

The temporal variation in meltwater contribution is evaluated using the results from the spectral analysis. Individual months characterized by pronounced meltwater recharge are identified by prominent peaks at a period of one day in fig. 4.11 and 4.12. Several authors report that after the cessation of snow melt, diurnal variations are pronounced at active rock glacier catchments exhibiting cirque glaciers, while lacking in those where cirque glaciers are absent (Potter, 1972; Tenthorey, 1993; Krainer and Mostler, 2002; Wagner et al., 2019a). The analysis presented above quantifies this effect. The fact that a pronounced impact is recognizable until early September, in one case even until October (discharge fluctuation 2014 due to a long, warm autumn), emphasizes the impact of cirque glacier presence within the catchment, providing an important source of water during the summer months, preserving diurnal discharge variations after snow melt has ceased and increasing the total annual runoff volume. This is in good agreement with the significant ice melt contribution predicted to last until early autumn by the rainfall runoff model (fig. 3.7, 5.1).

While melting rates at the glacier surface reflect the diurnal variation in radiation, heat and vapour content of the adjacent air (Cuffey and Paterson, 2010), the coarse grained active layer covering the rock glacier permafrost ice protects it from radiation and induces a damped and retarded variation in temperature (Haeberli, 1985; Vonder Mühll, 1993; Bevilacqua, 2019; Wagner et al., 2019a,b). The insulation and cooling provided by the blocky surface layer thermally decouples the permafrost ice to some extent from the external weather and climate conditions (Barsch, 1996; Jones et al., 2019; Wagner et al., 2019b). Accordingly,

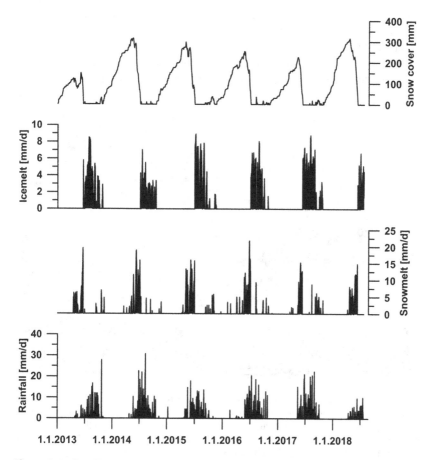

Figure 5.1 Specific recharge [mm/d] of rainfall, snow melt, and ice melt along with specific snow cover storage [mm] obtained from the rainfall runoff model of the Innere Ölgrube employed by Wagner et al. (in prep.)

melting of permafrost ice is expected to follow changes in atmospheric conditions in a damped and retarded fashion compared to melting of cirque glacier ice. Recharge-dominated spring responses reflect the input signal, i. e. the temporal distribution of meltwater production at the source. The distinct diurnal variation in spring discharge and EC evident from the overall hydrograph (fig. 3.2a), from the results of the spectral analysis (fig. 4.10–4.12), from the high significance

and periodic variation of the cross correlation coefficient between air temperature and discharge (fig. 4.13; cf. fig. 3.6c), as well as from the distinct periodic
variation of the cross correlation coefficient between discharge and EC (fig. 4.13;
cf. fig. 3.6c) point towards a distinct diurnal variation in meltwater production.
These results suggest the cirque glacier as major source of ice melt.

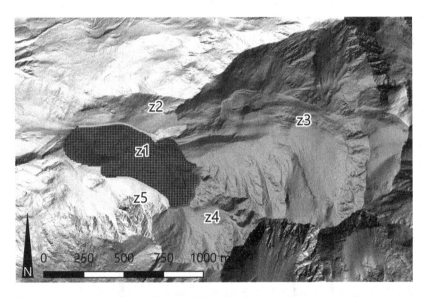

Figure 5.2 Recharge zones. The model domain of a prospective numerical groundwater
flow model is depicted as structured regular grid exhibiting a cell size of 10 x 10 m to facilitate comparison with Pauritsch et al. (2017). Peripheral cells affected by allogenic recharge
are colored appropriately along the rock glacier margin (digital elevation model: Open Data
Österreich, https://www.data.gv.at, CC BY 4.0)

The isotopic signature of meltwater derived from glaciers is similar to that
from melting permafrost ice within the rock glacier, thus separating the discharge
contributions from cirque glacier and rock glacier ice melt based on stable isotopes is not straightforward (Winkler et al., 2018b). However, several studies
comparing glacial meltwater to rock glacier outflow record strongly elevated ion
concentrations within the latter (see reviews by Colombo et al., 2018b, and Jones
et al., 2019). Williams et al. (2006) and Colombo et al. (2018a) combined water
chemistry and isotope studies to show that rock glacier (permafrost) ice exhibits

high dissolved ion contents and, consequently, its meltwater shows high electrical conductivity (EC) compared to cirque glacier meltwater. Harris et al. (1994) compared the runoff from a glacier to that from a nearby rock glacier, concluding that the latter exhibited an EC exceeding that of the former by a factor of 2. Melting of ice samples obtained by core drilling through the Gruben Rock Glacier exhibited high EC compared to glacial ice (Barsch, 1996). The increased ion content is attributed to the large mineral surface area within the ice-debris mixture of the frozen rock glacier core, promoting chemical weathering. Actively deforming rock glaciers contain a significant amount of newly eroded mineral surfaces, facilitating increased solute concentrations in the water content of the permafrost core that progressively increases during the summer season (note that due to the melting point depression, liquid water is already present in the rock glacier core at temperatures below 0°C, cf. Kurylyk and Watanabe, 2013). In addition, permafrost ice has been found to contain high ion concentrations that are not attributable to chemical weathering (Krainer et al., 2015). In summary, large amounts of glacial meltwater contributing to spring discharge should be recognized by lowering the EC due to dilution. In contrast, large amounts of permafrost ice meltwater is unlikely to lower the EC strongly (may not affect it at all or even raise it in some cases).

Since the rock glacier spring response falls into the recharge-dominated regime, melting processes driven by diurnal variation in air temperature are identified using cross correlation analysis. The time lag between positions of maximum equivalence between the air temperature and the discharge record roughly indicates the travel time of the hydraulic pulse. The EC of the corresponding meltwater is evaluated by correlating the discharge and EC time series accordingly. The strong *negative* correlation between discharge and EC during a dry period in early August 2017 (fig. 4.13a,b; cf. fig. 3.4 and 3.6), when snow melt had already ceased (cf. fig. 3.7, 5.1) indicates periodic diurnal dilution with low mineralized meltwater. This meltwater thus is probably attributable to cirque glacier ice melt – in case of permafrost ice melt within the rock glacier a positive correlation between EC and discharge (or no correlation at all) would be expected due to the presumably high ion concentration in permafrost ice melt water. Note, however, that the permafrost ice at the OEG has not been sampled yet, thus this statement represents an inherently uncertain conclusion by analogy rather than empirical evidence. The cross correlation results presented in fig. 4.14 and 4.15 and tab. 4.4 quantitatively show the general inverse correlation between discharge and EC, suggesting that this conclusion might be generalized at least to some degree (low mineralized cirque glacier meltwater causes dilution during increased discharge). The sinusoidal shape of the cross correlograms depicting

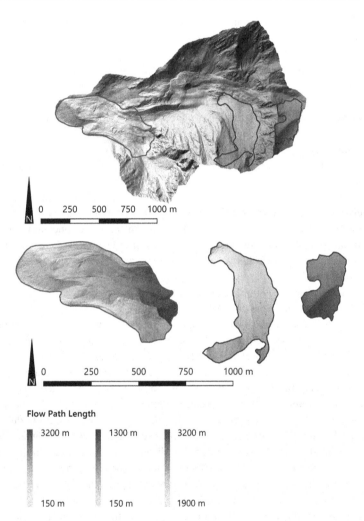

Figure 5.3 Flow path length distributions of the complete catchment, the rock glacier (red), and the cirque glaciers (blue) (digital elevation model: Open Data Österreich, https://www. data.gv.at,CC BY 4.0; glacier inventory: Buckel an Otto (2018): The Austrian Glacier Inventory GI 4 (2015) in ArcGis (shapefile) format., https://doi.pangaea.de/10.1594/PANGAEA. 887415, CC BY 3.0; rock glacier inventory: Wagner et al., 2020c, https://doi.pangaea.de/10. 1594/PANGAEA.921629, CC BY 4.0)

short-term variations in discharge and EC further substantiates conclusion, since after the cessation of snow melt diurnal periodic variations are attributable to cirque glacier melt (see above).

Glacial and permafrost ice melting processes occur at different locations within the rock glacier and/or its associated catchment. For melting of permafrost ice taking place within the rock glacier, the flow distance to the gauging station ranges from ~150 to ~1300 m, depending on the exact location of the meltwater production (fig. 5.3). Contrastingly, the flow distance to melting cirque glacier ice ranges from ~1900 to ~3200 m (fig. 5.3). The cross correlation results indicate that the hydraulic pulse arising from cirque glacier melting processes arrives with a retardation of 16 h at the spring (fig. 4.13c,d; cf. fig. 3.6c), allowing for a lower-bound estimate of its flow velocity: The shortest distance between the lowermost point of the cirque glacier and the gauging station is approximately 1900 m (fig. 5.3). However, the actual flow distance might exceed this value (to an unknown extent). Typically, runoff from glaciers reaches a maximum a few hours after the peak in melt, with decreasing difference as summer advances (Cuffey and Paterson, 2010). Thus, the peak in discharge at the glacier front will precede the peak at the gauge by less than 16 h. Both estimates suggest that the water travels from the front of the glacier to the gauging station at a velocity $\geq 3.3 \cdot 10^{-2}$ m s^{-1}. This is in good agreement with results from artificial tracer tests characterizing the fast flow component within the rock glacier, yielding slightly higher linear velocities (tab. 4.3; except for the method of moments which is heavily influenced by the tailing, see above).

In summary, these results identify glacial meltwater travelling at a linear velocity on the order of ~10^{-2} m s^{-1} as the causal agent of periodic diurnal variations in discharge and EC after the cessation of snow melt. This is in good agreement with results obtained by Colombo et al. (2018a) who showed that solute export from an active rock glacier lacking a contributing cirque glacier is driven primarily by rainfall. In contrast to air temperature and radiation, rainfall does not exhibit a distinct periodic diurnal variation, suggesting that if melting of permafrost ice is taking place it will follow a more irregular pattern (superimposed on the general seasonal trend). The good agreement between the period dominated by diurnal variations (fig. 4.11, 4.12) and the season dominated by ice melt according to the rainfall runoff model (fig. 3.7, 5.1) suggests that glacial meltwater accounts for most of the ice melt contribution. Note, however, that the methods presented do not account for long-term variations in recharge fluxes. Thus, an additional component provided by melting of rock glacier permafrost ice cannot be ruled out. Nevertheless, its contribution is probably comparatively small, as previously suggested by Croce and Milana (2002), Krainer and Mostler

(2002), Winkler et al. (2018b), and Wagner et al. (2019a). As a first approximation, therefore, the ice melt recharge component inferred from the rainfall runoff
model (fig. 3.7, 5.1) might be attributed solely to subcatchment z3 (fig. 5.2),
infiltrating into the rock glacier rooting zone as small creek and as distributed
groundwater flux.

In contrast, Heigert (2018) concludes that melting of permafrost ice within the
rock glacier accounts for a significant contribution to spring discharge. This view
is based on the EC and stable isotope trends depicted in fig. 3.5. The rock glacier
springs break the trend of increasing EC, δ^2H and δ^{18}O with decreasing altitude
and display lower variation than some of the catchment samples located above
the rock glacier rooting zone (see fig. 2.2b for sample locations; samples are
numbered in order of decreasing altitude). Fig. 5.4 shows the temporal evolution
of the analyzed natural tracers in all samples. Note contrasting distributions in
August and September: The spread of δ^2H and δ^{18}O values decreases in September, reflecting the convergence of isotopic signatures as groundwater becomes
the dominant source (fig. 3.6b; Heigert, 2018). In contrast, the spread in EC values recorded on 15.9.2017 strongly increases. Specifically, the high EC of creek
water immediately before infiltration into the rock glacier recorded in September
2017 exceeds the EC of rock glacier spring water by a factor of ~2. This thesis
attributes the contrasting behavior to a drop in temperature at the beginning of
September accompanied by snow fall and decreasing event water contributions
to spring flow (fig. 5.1, 5.4). Accordingly, diurnal discharge variations are attenuated at the beginning of September (fig. 5.4, 5.5a,b) and the cross correlation
between air temperature and discharge becomes insignificant (fig. 5.5c,d). Both
observations indicate the end of the meltwater dominated period after snowfall
has shut down melting of cirque glacier ice, reflected by low discharge of partly
frozen streamlets observed in the catchment on 15.9.2017 (Heigert, 2018).

Heigert (2018) explains these results by mixing with water exhibiting low EC
and strong depletion in ^{18}O inside the rock glacier. He suggests melting rock
glacier (permafrost) ice to be a plausible source of this water, acknowledging the
uncertainties associated with this conclusion. Note, however, that this suggestion
does not take the storage dynamics of the rock glacier into account and is in con-
flict with high EC values of permafrost ice melt reported by several authors (see
above). The results presented in this thesis support the conclusion by Heigert
(2018) that mixing with water exhibiting low EC and strong depletion in ^{18}O
inside the rock glacier is taking place. However, instead of attributing the source
of this water to permafrost ice within the rock glacier, it is suggested here that
the mixing involves low-EC and δ^{18}O-depleted water stored temporarily within
the rock glacier, stemming from rainfall and cirque glacier melt during August

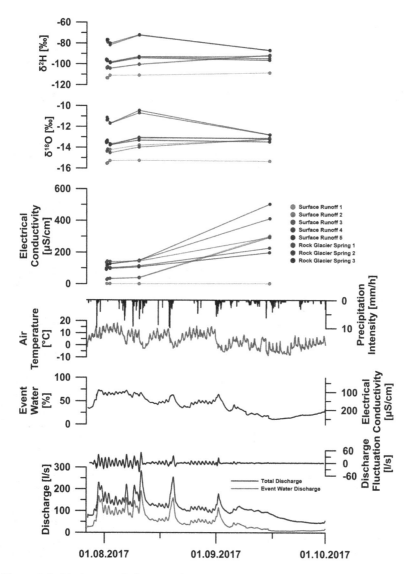

Figure 5.4 Discharge and discharge fluctuation, results of the two component mixing model (Heigert, 2018; Wagner et al., 2019a), air temperature and precipitation intensity during summer 2017. Evolution of EC and stable isotope signature in the catchment and rock glacier spring samples (locations outlined in fig. 2.2b; cf. fig. 3.4, 3.5)

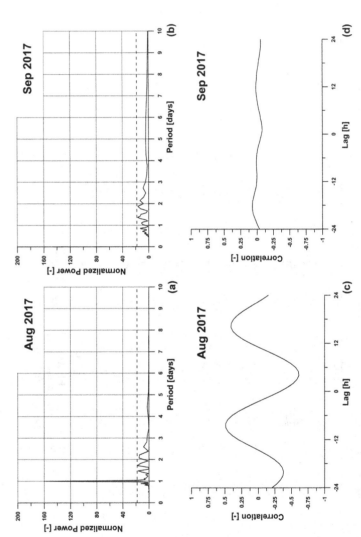

Figure 5.5 Spectral analysis of discharge fluctuations (a, b) and cross correlation between discharge fluctuation and air temperature fluctuation (c, d) computed for August and September 2017. Note the highly significant peak at a period of 1 day during August, while no significant peak can be observed in September. The cross correlation between air temperature and discharge shows a time lag of 16 h in August, while no significant correlation can be obtained for September 2017. These results are explained by the cold weather from the beginning of September, shutting down melting of the cirque glacier

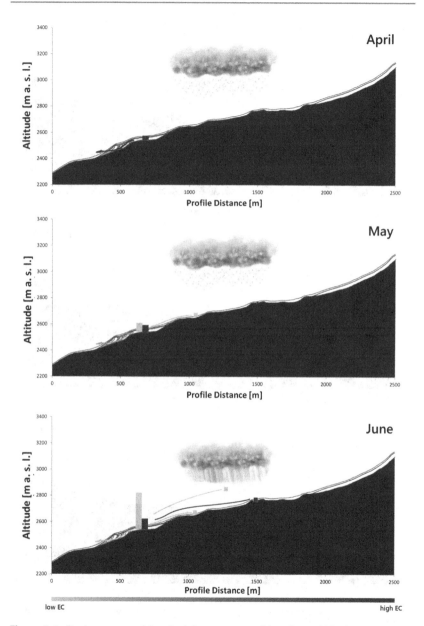

Figure 5.6 Dual storage model and mixing processes taking place within the rock glacier. The two reservoirs are depicted as light blue (event water store, low EC) and dark blue (ground water store, high EC)

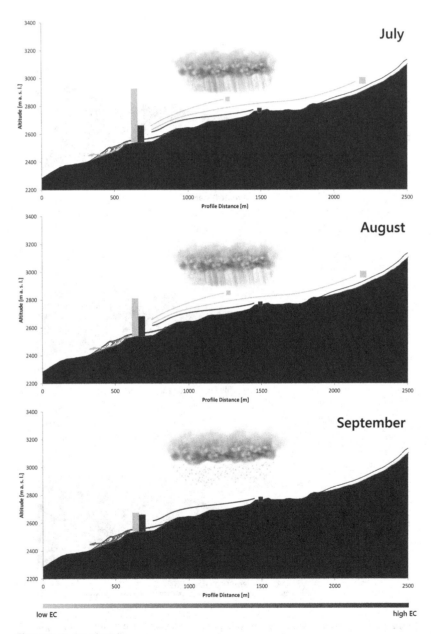

Figure 5.6   (continued)

Figure 5.6 (continued)

(fig. 5.6). This is in good agreement with the threshold analysis (Wagner et al., 2020b) demonstrating that a considerable amount of event water is temporarily stored (most of the water before the threshold is exceeded, 40 % after the threshold is exceeded) and released delayed at the spring. Similarly, the tracer tests results indicating an immobile zone accounting for 30 – 45 % of the fast component suggest considerable storage of water along the flow path of the fast flow component (in addition to the storage of large amounts of groundwater in the base layer). The two reservoirs are depicted as light blue (event water store, low EC) and dark blue (ground water store = base layer, high EC) in fig. 5.6. The event water store is provided by the heterogeneous structure of the permafrost table (providing swells and depressions for retaining some event water, fig. 4.3), by fine-grained domains within the active layer, and by the degrading permafrost close to the rock glacier front (increases storage of the permafrost layer). The EC samples Surface Runoff 2–5 taken in September are dominated by groundwater (high EC), since glacial melt (low EC) stopped by the end of August 2017 (fig. 3.4, 3.5, 3.6, 5.4, 5.5, 5.6; sample locations are indicated in fig. 2.2b). However, glacial meltwater from the preceding hot weeks during August (fig. 5.4, 5.5) is still stored in the rock glacier, diluting the highly mineralized recharge measured at Surface Runoff 4 and Surface Runoff 5. Thus, combining the mixing process assumed by Heigert (2018) with the threshold analysis of Wagner et al. (2020b), the moderate EC at the rock glacier spring is plausibly attributed to the dual rock glacier storage (fig. 5.6) instead of melting of permafrost ice inside the rock glacier. Similar observations at the Becs de Bosson Rock Glacier (Pennine Alps, Switzerland)[1] are also explained by storage within the rock glacier (Tenthorey, 1993).

The proposed model (mixing with temporarily stored low-EC water) is illustrated schematically in fig. 5.6. Before air temperatures increase above 0°C and snow begins to melt around April recharge is negligible, and (minimum) spring flow is sustained only by groundwater stored within the rock glacier, characterized by high EC. At the beginning of snow melt, the active layer has not fully developed and a thermokarst channel system enabling preferential flow paths is not available yet. Instead, water stemming from snow melt infiltrates diffusively into the rock glacier, increasing the hydraulic gradient within the unfrozen base layer. As a consequence, high-EC groundwater is pushed out, increasing the EC recorded during the first days after the beginning of snow melt (Heigert, 2018; Wagner et al., 2019a).

During the following weeks, snow melt reaches its maximum and rainfall provides additional recharge, filling the rock glacier storage until its maximum is reached in June or early July. At this point the active layer is partially thawed and the high amount of stored low-EC water is reflected in maximum spring flow (up to several hundred l/s) and minimum ground water contribution (~20 %) (fig. 3.2, 3.4). Note that it is not the amount of water stored but the amount of water released that determines the spring flow EC (i. e. even when the ground water store provided by the base layer is well filled, the higher permeability of the event water store (draining via preferential flow paths) dominates the spring flow. The spring water chemistry reflects the flux-averaged EC of the rock glacier storage rather than the volume-averaged EC). As pointed out by Heigert (2018), after the cessation of snow melt the maximum runoff decreases within days to values between 100 and 200 l s^{-1} (cf. fig. 3.2, 3.4). Although heavy summer thunderstorms cause sharp peaks superimposed on the seasonal pattern, only long lasting rainfall events (e. g. at the end of July in 2014, cf. fig. 3.2a) delay the overall recession trend significantly (Heigert, 2018). This indicates that in addition to the base layer (providing the major storage reservoir, residence time up to months, releasing correspondingly high-EC groundwater) a second reservoir characterized by lower residence time is filled with low-EC water. Water stored in this second reservoir is released within days or weeks. As the groundwater contribution steadily increases while discharge decreases during late summer, the short-term store is continuously recharged by glacial meltwater and rainfall events

---

[1] Exact location given in fig. 5.10 and tab. 5.2

but its internal water balance is negative due to its high permeability and drainage via preferential flow paths – while the groundwater store retains its contribution, resulting in an overall increase of the relative groundwater share (fig. 3.4; Heigert, 2018). However, these two reservoirs also interact (presumably via diffuse infiltration), as demonstrated by the two component mixing model (fig. 3.4): While event water prevails during peaks in response to summer thunderstorms, after the passage of the peak the groundwater contribution at the spring increases above its pre-event level, i. e. the recharge pulse hits also the long-term store of the rock glacier (Heigert, 2018). Water percolates through the frozen rock glacier core in the voids between the larger clasts (cf. Colombo et al., 2018a; Mohammed et al., 2018). Both the hydraulic interaction and the deep position of the 0°C isotherm within the rock glacier dampen the spring water EC response to precipitation events (fig. 3.4).

When air temperatures drop below the melting point, cirque glacier melt ceases and precipitation falls increasingly as snow, initiating a pronounced change in spring flow: daily variations are attenuated, ground water contribution increases, and singular rainfall events heavily impact the event water share (fig. 3.4). The reason is that the short-term reservoir runs dry and refreezes, thus the available short-term storage is lost and event water is no longer retarded, consequently affecting the spring water EC immediately (fig. 3.4, 5.6). Finally the short-term storage runs completely dry and base flow is sustained by groundwater from the base layer only (winter recession; fig. 5.6).

This thesis adapts the conceptual model of rock glacier hydrology proposed by Winkler et al. (2018a,b) and Wagner et al. (2019a), broadly dividing flow paths within the active rock glacier in supra-permafrost and sub-permafrost flow. This requires the frozen rock glacier core to exhibit comparatively low permeability in order to effectively separate these flow paths. In domains characterized by high volumetric ice content, interstitial ice seals the pores effectively, or massive ice lenses block groundwater flow (Krainer and Mostler, 2002; Hayashi, 2020). In parts of the rock glacier where water infiltrates into gaps and fractures and percolates through the permafrost body, however, the frozen rock glacier core cannot be regarded as a strict hydraulic barrier (Wagner et al., 2020b). While supra- and sub-permafrost flow are attributed to the active layer and the base layer, respectively, flow paths within the permafrost layer might either contribute to the fast flow component (preferential flow paths) or to the slow flow component (diffuse infiltration) identified by Winkler et al. (2018a). Hydraulic conductivity estimates of active layer, permafrost layer, and base layer derived in this thesis are presented in fig. 5.7. Note that the latter two actually represent lower bound estimates. The

conceptual model for the OEG, modifying the general model provided by Winkler
et al. (2018a,b) to account for site-specific characteristics, is depicted in fig. 5.8.

Figure 5.7 Hydraulic
conductivity estimates of
the active layer, permafrost
layer, and base layer. Note
that the latter two actually
represent lower bound
estimates. Numbers next to
boxplots indicate sample
size

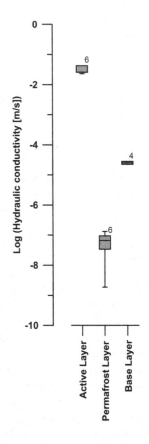

Supra-permafrost flow is associated with a seasonal shallow aquifer perched
on top of the permanently frozen rock glacier core. The dominance of coarse
debris and the absence of interstitial ice during the summer months constitute
a thin, highly permeable aquifer. As such, its base topography and permeability (corresponding to the permafrost table) is expected to govern flow conditions
within the shallow aquifer. The correlation of the active layer geometry to surface topography gives rise to enclosed depressions and linear features controlling
the heterogeneous flow field (fig. 4.3). The irregular distribution of frozen matrix
results in a highly heterogeneous structure, including a network of convoluted

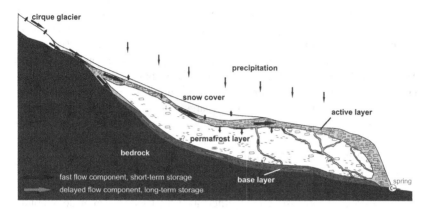

Figure 5.8 Conceptual Model (modified after Winkler et al., 2018b)

flow paths, depressions promoting storage in subsurface pools or thermokarst features, as well as crevasses and taliks acting as local swallets. High flow velocities are promoted in the coarse grained domains of the active layer, provided either by local erosion of finer particles (e. g. along preferential flow paths; Vick, 1981) or by saturation of the fine grained horizon underlying the blocky surface layer (Rogger et al., 2017). Linear flow velocities on the order of ~10^{-2} m s^{-1} are indicated by the results of two artificial tracer tests, reaching the nonlinear flow regime along preferential flow paths. The presence of preferential flow paths is substantiated by the high hydraulic conductivity (on the order of ~10^{-2} m s^{-1}) exceeding the expected hydraulic conductivity based on grain size analysis (silty sand and sandy silt), advective transport dominating over diffusive transport (high Peclet numbers), and the recharge-dominated spring response (indicated by the results from spectral analysis and cross correlation analysis) precluding the high response times associated with a homogeneous porous medium (several orders of magnitude too high; Covington et al., 2009). These flow paths provide the fast flow component identified by recession analysis and the two component mixing model (Heigert, 2018; Winkler et al., 2018b; Wagner et al., 2019a, 2020b). Along these flow paths a highly heterogeneous flow field is indicated by the pronounced tailing of the tracer breakthrough curve (fig. 4.6–4.8, indicating at least two domains of strongly contrasting flow velocities) as well as by the threshold analysis (fig. 3.3b; indicating the presence of storage zones and retardation of 40 % event water associated with the fast flow component).

The very fact that the rock glacier spring response is recharge-dominated implies that aquifer characterization based on spring hydrograph evaluation is of limited use for characterizing the properties of the channel system (Rehrl and Birk, 2010). Note that this does not affect the winter recession analysis outlined above, since the slow flow component response falls into the geometry dominated regime (as proven by the sustained base flow during the winter period lasting for several months without significant recharge). However, eq. 3.26 predicts that characteristic response time of the channel system is also below one day, i. e. the channel system is sufficiently simple to exert negligible influence on the passing meltwater recharge pulses. Most likely, this points towards a system of closed or open conduits, rather than a series of reservoirs and constrictions (Covington et al., 2009, 2012). This conclusion agrees well with the artificial tracer test results indicating a low dispersion coefficient. If the karst system analogy holds, the observed tailing of tracer load curves further substantiates this conclusion by precluding a complex series of constrictions and reservoirs along the flow path, which would reduce tailing but increase the overall dispersion (Jeannin and Maréchal, 1998; Hauns et al., 2001).

Percolation pathways of infiltrating water through the unsaturated zone are short before reaching the perched aquifer on top of the permanently frozen rock glacier core. A high portion of summer rainfall infiltrating into active rock glaciers generally runs off on top of the frozen rock glacier core. Infiltration into the highly permeable, coarse grained active layer bare of vegetation typically exceeds infiltration into deeper layers where permeability is reduced due to the presence of interstitial or massive ice (Krainer and Mostler, 2002). Water running off along the permafrost table is indicated by field observations of meltwater channels eroded into the frozen rock glacier core which are traceable (visible, audible) over distances of some tens of meters (Berger et al., 2004). While these concentrate along longitudinal furrows in the upper and middle part of the rock glacier, they are absent close to the rock glacier front, suggesting the water to infiltrate into the permafrost layer, subsequently flowing along englacial and subglacial meltwater tunnels towards the springs (cf. Krainer and Mostler, 2002). The short residence time (hours to days) of water within the active layer and/or permafrost layer (short-term storage, cf. fig. 5.6) explains the low mineralization of event water (making up ~60 % during the summer months) and the inverse correlation between discharge and EC. This short-term storage reservoir runs dry quickly inducing the fast early-time recession (fig. 3.3a; Wagner et al., 2020b), the isotope convergence in September (fig. 3.6b, 5.4) and the small soil moisture accounting store in the rainfall runoff model (Wagner et al., in prep.). Along with the blocky surface layer and the low temperatures in the alpine catchment, this invokes low evapotranspiration rates.

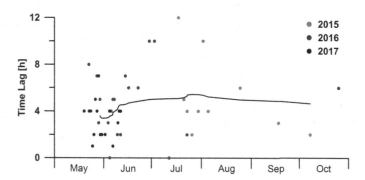

Figure 5.9 Time lags between discharge responses and subsequent EC responses to recharge events analyzed by Heigert (2018)

Supra-permafrost flow is heavily affected by seasonally changing meteorological conditions and local weather conditions influencing the thermal conditions in the active layer. Cold weather periods in summer or extended warm periods until late autumn heavily affect the spring flow pattern (fig. 3.2, 3.4, 5.4; substantiated by spectral analysis and cross correlation results). During the winter months the active layer is frozen. These seasonally changing patterns result in seasonally changing storage capacities of the short-term storage reservoir (negligible during winter and early spring, relatively high in summer). Accordingly, the preferential flow paths draining this reservoir carry different amounts of water depending on the season (high volumes during summer, comparatively low volumes during spring and autumn). This is reflected in the time lags between discharge responses and subsequent EC responses to recharge events analyzed by Heigert (2018). During the summer months, these show generally higher values, corresponding to larger volumes of water filling the preferential flow paths (fig. 5.9). The strong scatter observed during the summer months might indicate

(1) the highly variable fill-levels of the short-term store, depending on the time elapsed since the last precipitation event, and/or
(2) highly variable flow geometry due to enhanced rock glacier creep during the summer months.

The permanently frozen rock glacier core is characterized by a heterogeneous internal structure attributed to spatially and temporally differing thermal conditions at the permafrost table and the development of thermokarst features driven by advective heat transport (flowing water) into the rock glacier. These differences are driven by strong altitude gradients along the rock glacier and shading of the rock glacier rooting zone by the rugged ridge bounding the catchment to the south. The permafrost index distribution across the rock glacier indicates favorable conditions for permafrost in the rock glacier rooting zone, while the front of the rock glacier and part along its northern boundary have reached or already crossed the lower permafrost boundary (fig. 2.3c; Boeckli et al., 2012a,b). Continuous permafrost prevailing in the rooting zone is substantiated by the presence of three meltwater lakes, indicating an irregularly shaped and relatively impermeable permafrost table. Accordingly a perched aquifer is formed on top of the permafrost table, presumably promoted by the high ice content close to the permafrost table reported from many rock glaciers (see below). The high permeability contrast between the uppermost part of the permafrost layer and the highly permeable active layer above promotes the development of preferential flow paths along the permafrost table (thermokarst features). The depressions and pools identified by the threshold analysis outlined above (fill-spill-drain hypothesis; Wagner et al., 2020b) act as temporary storage zones for water flowing on top of the frozen permafrost body (promoted by high permeability contrast; threshold clearly observable in active rock glaciers (Wagner et al., 2020b), somewhat lower in an inactive rock glacier (Harrington et al., 2018), and poorly defined in the relict Schöneben rock glacier (Wagner et al., 2020b)). Evaluation of water level fluctuations of these meltwater lakes suggest that some water percolates through the permafrost body, exhibiting a hydraulic conductivity $> 10^{-8}$ m s^{-1}, that is probably several orders of magnitude higher in domains of the rock glacier characterized by degrading permafrost. Degrading permafrost prevails close to the rock glacier front, indicated by downslope decreasing ice content indicated by geophysical measurements (Hausmann et al., 2012), seasonal varying deformation rates in response to infiltrating water (Krainer and Mostler, 2006), and thickening of the rock glacier towards the front (bedrock slope angle exceeds surface slope angle; fig. 3.1a), although considerable amounts of ice are still present as indicated by the steep rock glacier front and geophysical measurements. The proximity of the lower permafrost boundary indicated by the permafrost distribution map (fig. 2.3c; Boeckli et al., 2012a,b) and the moderate Bouguer anomaly close to the rock glacier front (compared to the thick permafrost layer indicated by refraction seismics and GPR; compare fig. 2.3d to fig. 4.1) substantiate the permafrost degradation in this part of the rock glacier.

Degrading permafrost conditions close to the rock glacier front are further supported by a somewhat surprising result obtained from comparing the creep model results to the interpolated geophysical measurements available in this part of the rock glacier (fig. 4.4). Note that the geometrical model is actually based on the latter here. However, the creep model predicts *lower* permafrost thickness (up to 50 %) compared to the thickness derived from GPR and refraction seismics. In fact, neglecting processes such as meltwater infiltration or concentrated deformation along shear zones (i. e. attributing the corresponding increase in deformation rates to steady secondary creep only) suggests that the creep model should *overestimate* permafrost thickness in the lowermost part of the rock glacier. A possible explanation involves the employed creep exponent $n = 3$. While this value yielded acceptable results at the Galena Creek and Äußeres Hochebenkar Rock Glaciers (Konrad et al., 1999; Hartl et al., 2016), constant strain rate triaxial tests of frozen soil samples by Arenson and Springman (2005) indicate that n is actually a linear function of ice content (ranging from $n = 3$ for pure ice to $n = 0$ for absence of ice). Thus, in domains characterized by degraded permafrost the actual stress exponent might be lower than assumed in this thesis, resulting in the observed underestimate.

The presence of interstitial ice in the permafrost layer results in comparatively low permeability and low specific storage. Nevertheless, a hydraulic connection between the active layer (short-term storage of low-EC water) and the base layer (long-term storage of high-EC ground water) is indicated by mixing between event water and groundwater, requiring some transfer between these components. The low recovery rate of both artificial tracer tests (despite the generally low short-term storage in the active layer) substantiates these results. Increased EC of spring water in the first week after the beginning of snow melt and increased groundwater contributions after the hydraulic pulse of individual recharge events during the summer months hit the groundwater store indicate a hydraulic connection. This connection is characterized by a duality of flow components:

(1) Diffuse flow infiltrates areally distributed, locally increased by ponds, affects the deformation behavior (seasonal fluctuation) and contributes to the delayed flow component identified by Heigert (2018) and Wagner et al. (2020b).
(2) Concentrated flow contributing to the fast flow component infiltrates via preferential flow paths including macropores and thermokarst features, potentially also along shear zones, resulting in relatively rapid responses to precipitation events (increased hydraulic gradient).

Discontinuous permafrost promotes the development of intra-permafrost flow paths by lowering the breakthrough time of developing thermokarst channels. In combination with the extensional deformation field prevailing in the lower part of the rock glacier (fig. 2.3e,f; Krainer and Mostler, 2006; Hausmann et al., 2012), the development of crevasses, moulins or taliks acting as local swallets is promoted in this part of the rock glacier. In addition, the bedrock threshold and associated thinning of the permafrost layer recorded immediately above the lower-most part of the OEG is expected to locally facilitate the development of crevasses (by analogy to glacier mechanics, e. g. Hambrey and Lawson, 2000; Cuffey and Paterson, 2010; Colgan et al., 2016). If the permafrost layer acts as a barrier in the rock glacier rooting zone but is permeable near the rock glacier front, intra-permafrost flow subtracts water from the supra-permafrost flow above, potentially leading to a completely unsaturated active layer at the rock glacier snout. While present in the upper part of the rock glacier, water running along thermokarst channels is not observed in the lower part of the rock glacier (Berger et al., 2004), suggesting deeper flow paths. Preferential flow paths on top of the frozen permafrost body govern supra-permafrost flow in the upper parts of the rock glacier, while in the lower part some water might infiltrate into the permafrost layer along fractures or thermokarst conduits (i. e. close to the rock glacier front the permafrost layer is not regarded as a strict hydraulic barrier). The presence of such conduits may be typical for rock glaciers located close to the lower per-mafrost boundary (Arenson et al., 2010). The slightly observable multiple peaks revealed by the artificial tracer tests indicate a network of channels, similar to observations by Tenthorey (1992, 1993), Vonder Mühll et al. (2003) and Arenson et al. (2010). Detection of these features would require high-resolution refraction seismics (narrower geophone spacing compared to the design by Hausmann et al., 2012) in combination with electrical resistivity measurements as demonstrated by Hauck et al. (2011) and Mewes et al. (2017).

Intra-permafrost flow is affected by seasonally changing meteorological con-ditions as well as by the warming alpine climate. The short-term storage provided locally by depressions and crevasses of the permafrost table and the frozen rock glacier core underneath induces damping of extreme events during the summer (Heigert, 2018; Wagner et al., 2019b), but once freezing starts this ability reduces since less water is able to infiltrate into lower parts of the rock glacier (pro-nounced change in spring flow and event water share at the end of summer, cf. fig. 3.2b, 3.4, 5.4).

Sub-permafrost flow through the unfrozen base layer governs the long-term spring response and slow recession component sustaining base flow during the winter months (Wagner et al., 2020b). This flow path corresponds to the delayed flow component and serves as the rate-controlling hydrogeological unit during winter (Wagner et al., 2020b), since the hydraulic conductivity of the base layer is about 3 orders of magnitude lower than that of the fast flow component (on the order of 10^{-5} m s^{-1}, indicated by the long-term recession analysis). The relatively constant hydrochemistry of the rock glacier springs (fig. 3.5, 3.6, 5.4) indicates a large common reservoir within the rock glacier (Heigert, 2018), in good agreement with the large production store of the calibrated rainfall runoff model identified by Wagner et al. (in prep.). This is provided by the pore space (free of ice) of the fine-grained base layer. This storage contributes continuously to spring flow, allows for continuous mixing with event water (fig. 3.4, 3.6c) and produces the hysteresis observable in EC-δ^{18}O plots (fig. 3.6d; Heigert, 2018; Winkler et al., 2018b). This large storage dampens extreme precipitation events and sustains base flow during droughts (in contrast to the other flow paths it is fairly decoupled from external meteorological conditions), and presumably affects the downstream hydrological regime (cf. Wagner et al., 2016). The remarkable similarities of base flow among intact and relict rock glaciers (Wagner et al., 2020b) indicate that this storage is a feature of rock glacier independent from their state of activity. As pointed out by Pauritsch et al. (2017), this layer might as well be characterized by heterogeneity.

The OEG discharge is composed of at least two flow and storage components, a fast one characterized by high flow velocities (on the order of 10^{-2} m s^{-1}), short residence times (several hours) and low EC ($\ll$ 100 μS cm^{-1}), and a delayed one exhibiting low flow velocities (on the order of 10^{-5} m s^{-1}), long residence times (up to several months) and high EC ($\gg$ 100 μS cm^{-1}) (cf. Berger et al., 2004; Krainer et al., 2007; Heigert, 2018; Wagner et al., 2020b). A similar duality of storage and flow processes has been reported from many rock glaciers (Giardino et al., 1992; Krainer and Mostler, 2002; Krainer et al., 2007; Wagner et al., 2020b; see also reviews by Barsch, 1996; Burger et al., 1999, Winkler et al., 2018a, and Jones et al., 2019). In addition, OEG recharge is characterized by diffuse and concentrated components, the latter being provided by the small creek infiltrating in the rock glacier rooting zone. Similarly complex structures are frequently observed in karst systems, and define the alteration of recharge signals propagating through such aquifers (Geyer, 2008).

In summary, the deliverables of this thesis include

(1) the specified aquifer geometry as a set of digital elevation models ready for import into MODFLOW preprocessing software (e. g. ModelMuse). Steep sections might require corrections for numerical error (e. g. Ghost Node Correction Package, XT3D options, both available for MODFLOW 6).

(2) Quantified and tabulated recharge fluxes allowing for a specification of boundary conditions, ready for integration into a numerical groundwater flow model (e. g. Recharge Package, Well Package, Specified Head Package, all available for MODFLOW 6).

(3) An updated conceptual model based on Winkler et al. (2018b) and Wagner et al. (2019a). By accounting for the duality of recharge, storage, and discharge the spring flow and natural tracer response to diurnally and seasonally changing boundary conditions are explained in a consistent model.

The next section compares these results to existing rock glacier studies.

5.2 Comparing results to existing rock glacier studies

At the end of this section evidence from several active rock glaciers in mountainous regions is summarized. Generally, their characteristics are in good agreement with the hydrogeological features of the OEG described above, thereby increasing trust in the outlined conceptual model. In addition, they demonstrate that the OEG might be regarded as a representative active rock glacier and could therefore potentially serve as a 'benchmark test site' for future active rock glacier studies (as the Schöneben Rock Glacier does for relict rock glaciers). Fig. 5.10 shows the locations of the compared rock glaciers. Since different climatic settings are expected to affect the hydrology of the individual rock glaciers, they are compared in a standardized manner by depicting the rock glaciers together with the permafrost index map provided by Boeckli et al. (2012a,b), where available (fig. 5.11).

Many active rock glaciers display an internal structure similar to the OEG (Winkler et al., 2018a; Jones et al., 2019). The coarse surface layer free of fine particles is a typical feature of most rock glaciers. It is typically 2 to 5 m thick and exhibits a total porosity between 0.3 and 0.4 (Barsch, 1996). The components are arranged in a loose, unstable, edge-supported structure, with large void spaces in between (Burger et al., 1999). The lack of fine particles is attributed to erosion of finer particles by percolating water and frost heave of boulders (Barsch, 1996; Burger et al., 1999; Croce and Milana, 2002). The coarse grained, highly permeable structure and the absence of vegetation on active rock glaciers typically allow for rapid infiltration of water (Krainer and Mostler, 2002). Beneath that boulder layer, poorly sorted sediments dominated by sand and silt are found at the Gruben Rock Glacier (Barsch et al., 1979; Barsch, 1996) and Reichenkar Rock Glacier (Krainer et al., 2000), similar to observations at the OEG (Berger et al., 2004). Samples taken beneath the coarse surface layer at the snout of Cadin del Ghiacciaio Rock Glacier, Cadin di Croda Rossa Rock Glacier, and Murfreit Rock Glacier exhibit poorly sorted sediments dominated by sand and gravel (Krainer et al., 2010, 2012). Core drilling samples taken from Murtèl-Corvatsch Rock Glacier and Muragl Rock Glacier (Arenson et al., 2002) indicate well graded sandy gravel with silt, although it is difficult to estimate the maximum grain size due to the limited core diameter. At the Lazaun Rock Glacier, gravel, cobbles and boulders dominate the core drilling samples, while sand, silt and clay typically make up less than 1 % (locally, sand might constitute an important fraction of the grain size spectrum; Krainer et al., 2015). The absence of clay particles found in these samples is attributed to insufficient degradation or weathering at high altitudes (Arenson et al., 2002). Note that except for the drill core samples, all samples were taken at shallow depths, within or directly underneath the active layer. As pointed out by Arenson et al. (2002), finer particles that are washed out of the coarse surface layer during the summer months might accumulate in this zone, questioning the representativeness of these grain size analyses for the overall rock glacier core. All core drilling samples exhibit coarse grained sediments (gravel, sand, cobbles, boulders). These are assumed to be less biased than samples taken directly beneath the coarse grained surface layer. These are in good agreement with the inferred properties of the Schöneben relict rock glacier, where complete thawing of permafrost left a highly permeable debris accumulation behind, exhibiting a hydraulic conductivity on the order of 10^{-2} to 10^{-3} m s^{-1} corresponding to dominating grain sizes of gravel or coarser material (Winkler et al., 2016a; Pauritsch et al., 2017). Note however, that preferential flow paths might yield a highly permeable aquifer even when large domains are characterized by fine grained sediment. At the relict Hochreichart Rock Glacier, excavations up to 6.5 m at the front revealed sandy gravel, cobbles and boulders. Finer particles account for less than 20 % and are completely lacking along washed-out channels forming preferential flow paths (Untersweg and Proske, 1996; Pauritsch et al., 2017). Note that caution is necessary when comparing grain size distributions of active and relict rock glaciers, since the melting of permafrost ice during rock glacier collapse might wash out fine particles (Untersweg and Schwendt, 1995, 1996).

Active layer thickness is frequently found to decrease with increasing altitude (e. g. Vonder Mühll, 1993; Barsch, 1996; Croce and Milana, 2002). Being defined by thermal constraints, it generally follows the surface topography in an attenuated form. Specifically, the active layer is often thicker on transverse ridges but lower in the furrows characterizing many rock glacier surfaces (Barsch, 1996).

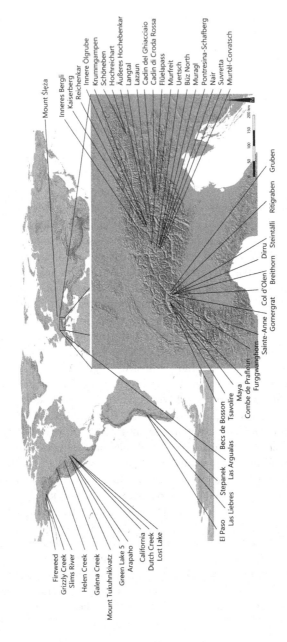

Figure 5.10 Locations of rock glaciers. Activity status, references, and coordinates are given in tab. 5.2 (digital elevation model: Copernicus Land Monitoring Service, 2015, http://land.copernicus.eu/pan-european/satellite-derived-products/eu-dem/eu-dem, funded by the European Union)

However, its local characteristics are controlled by air temperature and solar radiation, by the spatial and temporal distribution of the snow cover and its physical properties, by subsurface composition and meltwater content, as well as by surface characteristics and topography (Barsch, 1996; Luetschg et al., 2004, 2008; Zhang, 2005; Zenklusen Mutter and Philips, 2012; Buchli et al., 2013). Similar to many other rock glaciers, the OEG surface is characterized by a highly heterogeneous grain size distribution, causing a corresponding variability in subsurface thermal conductivity (Barsch, 1996; Wagner et al., 2019b). As a result, the active layer thickness is characterized by strong spatial and temporal variability. An increasing trend in active layer thickness recognized at many rock glaciers is attributed to global warming (Barsch, 1996).

The overall ice content of active rock glaciers has been estimated to range typically from 40 to 80 % of the total volume (Barsch, 1996). The volumetric ice content of the permanently frozen rock glacier core varies significantly, both within and among individual rock glaciers. Core drillings and geophysical investigations indicate ice contents ranging from 10 to 100 % (Arenson and Jakob, 2010). Two core drillings profiles at the Lazaun Rock Glacier indicate an ice content of 50 % in the middle part of the rock glacier, while permafrost degradation towards the front reduces it to 22 % (overall average 35–40 %; Krainer et al., 2015). Both profiles exhibit strongly varying ice content with depth (Krainer et al., 2015). Combined geophysical investigations at the lower parts of Reichenkar Rock Glacier indicate an average ice content of 45–60 %, displaying strong lateral variations (Hausmann et al., 2007), while its upper part is characterized by a massive ice core (Krainer et al., 2000). At Kaiserberg Rock Glacier, an average ice content of 40–50 % is inferred, again showing pronounced lateral variations (Hausmann et al., 2012). At the El Paso Rock Glacier, refraction seismics indicate an average ice content of 58 %, with permafrost deterioration progressing towards the front (Croce and Milana, 2002). Ice exposures and georadar measurements at the Murfreit Rock Glacier indicate a massive ice core in the upper part, while the lower part is characterized by discontinuous ice (Krainer et al., 2012). Similarly, geoelectrical measurements at the Las Argualas Rock Glacier indicate decreasing ice content with decreasing altitude (Fabre et al., 1995).

The frozen rock glacier core is made up of ice-cemented debris (which might be either undersaturated, saturated or supersaturated with ice), ice lenses of several meters thickness or a continuous massive ice core. Strong variations in ice content with depth are typical (Barsch, 1996; Arenson and Jakob, 2010). Meltwater flowing along the permafrost table is frequently observed, indicating the low permeability of the shallow frozen rock glacier core (Barsch et al., 1996). This zone immediately below the permafrost table is found to contain high amounts of ice in many cases (often ice-supersaturated debris or massive ice; Potter, 1972; Fisch et al., 1977; Haeberli and Patzelt, 1982; Haeberli, 1985; King et al., 1987; Haeberli et al., 1988; Vonder Mühll, 1993; Barsch, 1996; Arenson et al., 2002; Mewes et al., 2017).

Seasonal ponds developing on top of the permanently frozen rock glacier core have been frequently observed, e. g. at Murfreit Rock Glacier (Krainer et al., 2012), Cadin del Ghiacciaio Rock Glacier (Krainer et al., 2010) Gornergrat Rock Glacier (Haeberli, 1985), Furggwanghorn Rock Glacier (Springman et al., 2012; Buchli et al., 2018), and Gruben Rock Glacier (Haeberli et al., 2001; Kääb and Haeberli, 2001). Similar features are reported from several rock glaciers in the American Cordillera (Brown, 1925; Wahrhaftig and Cox, 1959; Johnson, 1978; Blumstengel and Harris, 1988; Giardino et al., 1992; Liaudat et al., 2020). Their presence indicates sufficiently low permeability of the underlying permafrost layer to inhibit immediate drainage into the deeper parts of the rock glacier.

As a consequence, water infiltrating into the active layer frequently flows on top of the frozen rock glacier core. This water is typically derived from rainfall, snow melt or melting cirque glaciers within the rock glacier catchment (if present) and frequently reaches the spring within a few hours after the recharge pulse (Krainer and Mostler, 2002). Highly heterogeneous flow fields within the active layer have been reported from a range of rock glaciers that are subject to climatic conditions similar to the OEG (fig. 5.11, tab. 5.1, 5.2). Water flowing concentrated along meltwater channels eroded into the permafrost table has been reported from the upper parts of Reichenkar rock glacier, Kaiserberg rock glacier and Langtal rock glacier, suddenly disappearing in the middle or lower parts of these rock glaciers (Krainer and Mostler, 2002). Similar observations are reported from the Muragl Rock Glacier, Murtèl-Corvatsch Rock Glacier (Vonder Mühll et al., 2003; Arenson et al., 2010; Springman et al., 2012), Furggwanghorn Rock Glacier (Buchli et al., 2013, 2018), Tsavolire Rock Glacier, Becs de Bosson Rock Glacier (Tenthorey, 1992, 1993), as well as from several rock glaciers in the American Cordillera (Wahrhaftig and Cox, 1959; Potter, 1969, 1972; White, 1971; Johnson, 1978; Giardino et al., 1992).

A permeable, heterogeneous active layer is indicated by several tracer tests exhibiting high flow velocities (several cm s^{-1}) and complex breakthrough curves showing multiple peaks in some cases (tab. 5.1). The strong heterogeneity is also reflected in the fact that despite close injection sites of two different tracers, only one of them was monitored downstream during several tests at various sites across the Furggwanghorn rock glacier (Buchli et al., 2013). In case of the Becs de Bosson rock glacier, injection at two sites within the rooting zone yielded a

Table 5.1 Tracer tests at intact rock glaciers

Flow velocity range [m s^{-1}]	Tracer	Rock Glacier	Assumed flow path	Remarks	Reference
$5.9 \cdot 10^{-2}$ – $9.1 \cdot 10^{-2}$	Uranine	Reichenkar	Supra-permafrost	Injection along meltwater currents eroded into the permafrost table; high velocities interpreted as channelized supra-permafrost flow along a network of conduits	Krainer and Mostler (2002)
$1.5 \cdot 10^{-2}$ – $3.5 \cdot 10^{-2}$	Uranine	Langtal	Supra-permafrost	Injection along meltwater currents eroded into the permafrost table; high velocities interpreted as channelized supra-permafrost flow along a network of conduits	Krainer and Mostler (2002)
$3.7 \cdot 10^{-2}$	Uranine	Kaiserberg	Supra-permafrost	Injection along meltwater currents eroded into the permafrost table; high velocities interpreted as channelized supra-permafrost flow along a network of conduits	Krainer and Mostler (2002)
$5.8 \cdot 10^{-2}$	Uranine	Tsavolire	Supra-permafrost	Injection in perennial snowfield meltwater current infiltrating in rock glacier rooting zone; high flow velocity and audible water currents below the blocky rock glacier surface layer interpreted as supra-permafrost flow; tracer injection: 20.8.1986	Tenthorey (1992, 1993)

(continued)

Table 5.1 (continued)

Flow velocity range [m s⁻¹]	Tracer	Rock Glacier	Assumed flow path	Remarks	Reference
$5.8 \cdot 10^{-2}$	Eosin	Tsavolire	Supra-permafrost	Injection in perennial snowfield meltwater current infiltrating in rock glacier rooting zone; single tracer breakthrough curve peak, high flow velocity and audible water currents below the blocky rock glacier surface layer interpreted as supra-permafrost flow; tracer injection: 8.8.1990	Tenthorey (1992, 1993)
$2.4 \cdot 10^{-5} - 2.9 \cdot 10^{-4}$	Uranine	Becs de Bosson	Sub-permafrost	Injection in perennial snowfield meltwater current infiltrating in rock glacier rooting zone; assumed flow path below permafrost layer; high uncertainty in flow velocity since recorded by charcoal bags only; tracer injection: 2.8.1986	Tenthorey (1992, 1993)
$3.8 \cdot 10^{-2} - 4.3 \cdot 10^{-2}$	NaCl	Becs de Bosson	Supra-permafrost (August); Sub-permafrost (September)	Injection in perennial snowfield meltwater current infiltrating in rock glacier rooting zone; complex tracer breakthrough curve displaying multiple peaks; high flow velocity and audible water currents below the blocky rock glacier surface layer interpreted as supra-permafrost flow; tracer injection on 18.8.1987 in meltwater current from perennial snowfield; repetition of test (same tracer, same injection and sampling locations) on 10.9.1987 did not show any response within 48 h, i.e. superficial flow supplies the sources only when the mean daily temperature is greater than zero	Tenthorey (1992, 1993)

(continued)

Table 5.1 (continued)

Flow velocity range [m s⁻¹]	Tracer	Rock Glacier	Assumed flow path	Remarks	Reference
$1.9 \cdot 10^{-4}$ – $3.9 \cdot 10^{-4}$	Uranine	Maya	Intra-permafrost; Sub-permafrost	Injection in perennial snowfield meltwater current infiltrating in rock glacier rooting zone; at the base of the active layer infiltration into the permafrost layer occurs; consequent transport through fine grained matrix and fissured bedrock; tracer injection: 2.8.1990	Tenthorey (1992, 1993)
$\geq 5 \cdot 10^{-3}$	Rhodamine G	Furggwanghorn	Supra-permafrost	water visible on top of permafrost table; nearby injection of uranine not recovered, indicating heterogeneous flow paths and poor hydraulic connections between adjacent locations; tracer injection: 23.8.2011	Buchli et al. (2013)
$1.2 \cdot 10^{-2}$	Rhodamine G	Furggwanghorn	Supra-permafrost	tracer injection on top of rock glacier front; rock glacier partially snow covered, indicating shallow 0°C isotherm; water visible on top of permafrost table; tracer injection: 10.7.2012	Buchli et al. (2013)
$9 \cdot 10^{-3}$	Uranine	Furggwanghorn	Supra-permafrost	tracer injection on top of rock glacier front; rock glacier partially snow covered, indicating shallow 0°C isotherm; water visible on top of permafrost table; water tracer injection: 10.7.2012	Buchli et al. (2013)

(continued)

Table 5.1 (continued)

Flow velocity range [m s⁻¹]	Tracer	Rock Glacier	Assumed flow path	Remarks	Reference
$6 \cdot 10^{-3}$	Rhodamine G	Furggwanghorn	Supra-permafrost	water visible on top of permafrost table; tracer injection: 29.8.2012; flow velocity low compared to earlier tracer tests (10.7.2012) despite steeper flow path; assumed to indicate greater depth of 0°C isotherm resulting in longer flow paths through thawed fine-grained material below blocky surface layer; nearby injection of uranine not recovered, indicating heterogeneous flow paths and poor hydraulic connections between adjacent locations	Buchli et al. (2013)
$4 \cdot 10^{-3}$	Rhodamine G	Furggwanghorn	Supra-permafrost	water visible on top of permafrost table; tracer injection: 29.8.2012; flow velocity low compared to earlier tracer tests (10.7.2012) despite steeper flow path; assumed to indicate greater depth of 0°C isotherm resulting in longer flow paths through thawed fine-grained material below blocky surface layer; nearby injection of uranine not recovered, indicating heterogeneous flow paths and poor hydraulic connections between adjacent locations	Buchli et al. (2013)
$1.2 \cdot 10^{-1}$ – $1.5 \cdot 10^{-1}$		Suvretta		creek blocked by rock glacier movement infiltrates into Suvretta Rock Glacier; neither tracer nor assumed flow path specified	Vonder Mühl (1993)

response within several hours in one case (attributed to supra-permafrost flow) and within several months in the other case (attributed to sub-permafrost flow; Tenthorey, 1992, 1993). Repeated implementation of the exactly same tracer test setting (supra-permafrost flow at Becs de Bosson rock glacier) yielded quick passage of the tracer during August (first test) but no recovery during September (second test indicating temporally varying flow field geometry (attributed to melting/freezing processes; Tenthorey, 1992, 1993). In all reported cases high flow velocities and channelized water currents are associated with meltwater currents from cirque glaciers or (perennial) snow fields infiltrating into the rock glacier rooting zone. In the absence of these features, diffuse water running along the permafrost table or infiltrating into the permafrost layer exhibits progressively reduced flow velocities during the summer (Buchli et al., 2013). This is attributed to deepening of the 0°C isotherm and corresponding increase in flow path length and/or ongoing thawing of fine grained material present below the blocky surface layer (Buchli et al., 2013). A similar process is described at Krummgampen rock glacier (Rogger et al., 2017).

Drill core samples taken from ice-rich permafrost layers exhibit air voids up to volumetric air contents of 25 % (Arenson, 2002; Arenson et al., 2004), providing pathways for circulating water within and below the permafrost body (Haeberli et al., 2006; Hausmann et al., 2007). Geophysical investigations indicate the presence of substantial amounts of water within the frozen rock glacier core of El Paso Rock Glacier (Croce and Milana, 2002). Core drillings at Furggwanghorn Rock Glacier, Murtèl-Corvatsch Rock Glacier, Muragl Rock Glacier and Pontresina-Schafberg Rock Glacier revealed the presence of water within the permafrost layer (Arenson et al., 2002; Zenklusen Mutter and Phillips, 2012; Buchli et al., 2013, 2018). At the Murtèl-Corvatsch Rock Glacier, the water content increases with decreasing altitude, probably because of the correspondingly increasing amount of meltwater transport at the permafrost table, and increasing temperature throughout the rock glacier core related to downslope movement and the warming climate (which affects the lower parts more than the shaded upper parts; Vonder Mühll, 1993; Barsch, 1996). At this rock glacier, runoff drains into a network of thermokarst channels within the permafrost body, which exhibit distinct properties depending on their depth (Vonder Mühll et al., 2003; Haeberli et al., 2006; Arenson et al., 2010). An intra-permafrost talik is indicated by thermal anomalies at the Furggwanghorn Rock Glacier (Zhou et al., 2015). In addition, water fills a depression at this rock glacier during the snow melt period, forming a thermokarst lake for a couple of weeks, until development of a drainage path (Buchli et al., 2018). Accelerated crevasse deepening in the Furggwanghorn rock glacier rooting zone (Roer et al., 2008) has also been attributed to thermokarst development (Springman et al., 2012). Intra-permafrost flow has been deduced from tracer test results at the Maya rock glacier (Tenthorey, 1992, 1993, 1994). Note that for all these rock glaciers, the permafrost index map predicts more continuous permafrost than at the OEG (fig. 5.11), substantiating the assumption of degrading permafrost and intra-permafrost flow occur at the OEG as well. Some of these examples are old compared to the permafrost distribution map of Boeckli et al. (2012a,b)—in these cases, permafrost was probably even less affected by degradation than suggested by the index map due to progressing global warming. Rapid acceleration of deformation rates is observed at the Büz North rock glacier during the snow melt period, stemming from the infiltration of meltwater into the permafrost body and associated reduction in effective stress (Ikeda et al., 2008). These observations are in good agreement with the reported seasonal acceleration of deformation rates in the lower part of OEG during the summer months, supporting the explanation by infiltrating meltwater given by Krainer and Mostler (2006; see above).

The contribution of melting permafrost ice to rock glacier spring discharge is typically considered to be low (e. g. Croce and Milana, 2002; Krainer and Mostler, 2002; Winkler et al., 2018a; Jones et al., 2019; Wagner et al., 2020b). At the Lazaun Rock Glacier an average permafrost ice melt contribution of 2.3 % of total discharge is estimated by Krainer et al. (2015). Specifically, the melting rate of permafrost ice in rock glaciers is low compared to the melting rate of glacier ice, because of the thermic insulation provided by the debris layer (Jones et al., 2019; Hayashi, 2020; Wagner et al., 2020b).

A fine grained, unfrozen sediment layer has been found at the base of several active rock glaciers, including Reichenkar Rock Glacier (Hausmann et al., 2007), Kaiserberg Rock Glacier (Hausmann et al., 2012), Lazaun Rock Glacier (Krainer et al., 2015), Murtèl-Corvatsch Rock Glacier (Haeberli, 1988), Muragl Rock Glacier (Arenson et al., 2002; Musil et al., 2006), Suvretta Rock Glacier (Vonder Mühll, 1993), Becs de Bosson Rock Glacier (Tenthorey, 1992, 1993), Maya Rock Glacier (Tenthorey, 1993), Murfreit Rock Glacier (Krainer et al., 2012), Sainte-Anne Rock Glacier (Evin, 1983), Vallecitos Rock Glacier (Corte, 1987), and El Paso Rock Glacier (Croce and Milana, 2002). As a consequence of the freezing point depression induced by overburden pressure, water ion content and/or pore water pressure, the base of frozen permafrost does not necessarily correspond to the $0°$ isotherm (Vonder Mühll, 1993). The unfrozen base layer is held responsible for the slow flow component, providing sufficient storage to maintain base flow during extended periods without recharge and buffering capabilities during high flow events (Tenthorey, 1992, 1993, 1994; Winkler et al., 2018a,b; Wagner et al., 2020b).

Since the presence of a permanently frozen rock glacier core profoundly influences groundwater percolation through active rock glaciers, their runoff pattern is expected to be markedly different from relict rock glaciers. However, combined analysis of the spring hydrograph, natural and artificial tracers and several refraction seismic profiles at the relict Schöneben Rock Glacier revealed remarkable similarities (Winkler et al., 2016a,b): Similar to active rock glaciers, a fast and a delayed flow component is found at the Schöneben rock glacier. The fast component is associated with a highly conductive, coarse-grained debris layer up to several tens of meters thick (former permafrost layer and active layer), controlling the fast and flashy spring responses to recharge events (Winkler et al., 2016a,b; Pauritsch et al., 2017). The delayed flow component is related to a fine-grained base layer of 15 m thickness at maximum, providing sufficient storage to sustain the base flow at the rock glacier spring (Winkler et al., 2016a,b; Pauritsch et al., 2017). Tracer tests (Winkler et al., 2016a,b), analytical models of sloping aquifers (Pauritsch et al., 2015; Winkler et al., 2016b) as well as various numerical groundwater flow model scenarios (Pauritsch et al., 2017) consistently suggest the hydraulic conductivity of this base layer to be on the order of 10^{-5} m s^{-1}, while on average the specific yield equals 0.3. Comparing the recession patterns of the Innere Ölgrube and Reichenkar Rock Glacier (both active) to the inactive Helen Creek Rock Glacier and the relict Schöneben Rock Glacier reveals a comparable overall behavior of all four rock glaciers (Wagner et al., 2020b). More specifically, slightly higher recession coefficients associated with the fast flow component are found at the active rock glaciers (Wagner et al., 2020b). These are attributed to lateral flow on top of the frozen rock glacier core, allowing for rapid drainage of the active rock glaciers compared to the thicker unsaturated zones found in inactive or relict rock glaciers (Winkler et al., 2016a,b; Pauritsch et al., 2017; Wagner et al., 2020b). However, the recession coefficients associated with the slow flow component are similar for all four rock glaciers, which is attributed to a fine-grained base layer exhibiting similar hydraulic properties in all four cases (Wagner et al., 2020b). This layer is assumed to exhibit significant storage capacity, providing base flow by releasing stored groundwater during winter (and during dry summer periods), while buffering extreme precipitation events (Wagner et al., 2020b).

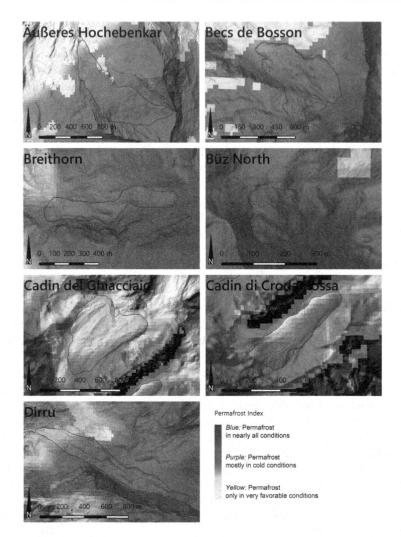

Figure 5.11 Studied rock glaciers. Likelihood of permafrost occurrence is given by Boeckli et al. (2012a,b). Rock glacier outlines in Austria are given by Wagner et al. (2020a). Rock glacier outlines in South Tyrol are given by Bollmann et al. (2012). Rock glacier outlines in Switzerland are drawn approximately according to King (1990), Tenthorey (1992, 1994), Vonder Mühll (1993), Barsch (1996), Kääb et al. (1997), Krainer et al. (2015), Buchli et al. (2018) (alpine permafrost index map: Boeckli et al., 2021b, https://doi.pangaea.de/ 10.1594/PANGAEA.784450, CC BY 4.0; digital elevation models: Copernicus Land Monitoring Service, 2015, http://land.copernicus.eu/pan-european/satellite-derived-products/eu-dem/eu-dem, funded by the European Union)

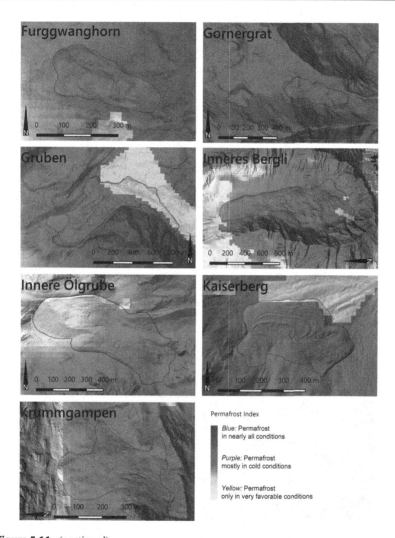

Figure 5.11 (continued)

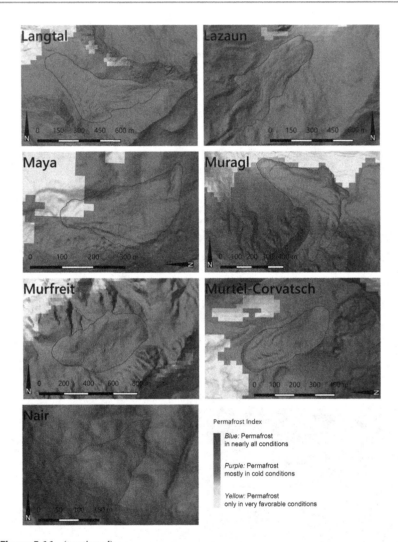

Figure 5.11 (continued)

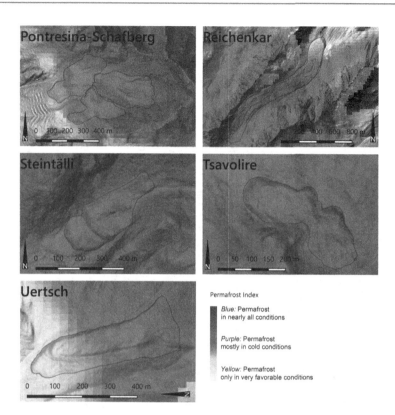

Figure 5.11 (continued)

Table 5.2 Rock glacier studies

Name	State	Mountain range	Position	Rock Glacier Inventory ID	References
Arapaho	active	Front Range, Colorado	105.640°W, 40.021°N		White (1971)
Äußeres Hochebenkar	active	Ötztal Alps, Austria	11.012°E, 46.832°N	LfCode 7RG1714 (Wagner et al., 2020a)	Haeberli and Patzelt (1982); Barsch (1996); Kääb et al. (2003); Thies et al. (2013); Klug et al. (2014); Graßmair and Erschbamer (2015); Hartl and Fischer (2015); Klug (2015); Krainer (2015); Hartl et al. (2016); Brodacz (2019); Wagner et al. (2020b)
Becs de Bosson	active	Pennine Alps, Switzerland	7.513°E, 46.172°N		Tenthorey (1992, 1993 1994); Mari et al. (2013); Mewes et al. (2017)
Breithorn	active	Pennine Alps, Switzerland	7.819 E, 46.139 N		Wirz et al. (2016)
Büz North	active	Albula Alps, Switzerland	9.818°E, 46.534°N		Ikeda and Matsuoka (2006); Ikeda et al. (2008)
Cadin del Ghiacciaio	active	Dolomites, Italy	12.149°E, 46.644°N	ID 1366 (Autonome Provinz Bozen, 2018)	Krainer et al. (2010)
Cadin di Croda Rossa	active	Dolomites, Italy	12.153°E, 46.637°N	ID 1094 (Autonome Provinz Bozen, 2018)	Krainer et al. (2010)
California	active	Sierra Blanca, Colorado	105.486°W, 37.617°N		Giardino et al. (1992)

(continued)

Table 5.2 (continued)

Name	State	Mountain range	Position	Rock Glacier Inventory ID	References
Col d'Olen	active	Pennine Alps, Switzerland	7.865°E, 45.868°N		Colombo et al. (2018a,c)
Combe de Prafleuri	active	Pennine Alps, Switzerland	7.378°E, 46.073°N		Fisch et al. (1977); Haeberli (1985); Barsch (1996)
Dirru	active	Pennine Alps, Switzerland	7.821 E, 46.120°N		Wirz et al. (2016)
Dutch Creek	active	Sierra Blanca, Colorado	105.450°W, 37.614°N		Giardino et al. (1992)
El Paso	active	Agua Negra range, Argentina	69.809°W, 30.229°S		Croce and Milana (2002)
Fireweed	active	Wrangell Mountains, Alaska	143.080°W, 61.469°N		Elconin and Chapelle (1997)
Flüelapass	active	Albula Alps, Switzerland	9.949°E, 46.746°N		King et al. (1987); Barsch (1996)
Furggwanghorn	active	Pennine Alps, Switzerland	7.729°E, 46.194°N		Roer et al. (2008); Springman et al. (2012); Buchli et al. (2013); Zhou et al. (2015)
Galena Creek	active	Absaroka Mountains, Wyoming	109.791°W, 44.644°N		Potter (1969, 1972); Konrad et al. (1999)

(continued)

Table 5.2 (continued)

Name	State	Mountain range	Position	Rock Glacier Inventory ID	References
Gornergrat	active	Pennine Alps, Switzerland	7.787°E, 45.986°N 7.780°E, 45.986°N		Haeberli (1985); King (1990)
Green Lake 5	active	Front Range, Colorado	105.628°W, 40.052°N		Liu et al. (2004); Williams et al. (2006); Leopold et al. (2011)
Grizzly Creek	active	St. Elias Mountains, Yukon	139.126°W, 61.085°N		Johnson (1978)
Gruben	active	Pennine Alps, Switzerland	7.971°E, 46.170°N		Barsch et al. (1979); Haeberli (1985); King et al. (1987) Barsch (1996); Kääb et al. (1997); Haeberli et al. (2001); Kääb and Haeberli (2001)
Helen Creek	inactive	Canadian Rockies, Alberta	116.415°W, 51.679°N		Harrington (2017); Harrington et al. (2018); Wagner et al. (2020b)
Hochreichart	relict	Seckau Alps, Austria	14.703°E, 47.371°N	LfCode 6RG65 (Wagner et al., 2020a)	Untersweg and Schwendt (1995, 1996); Untersweg and Proske (1996)
Innere Ölgrube (syn. Ölgrube Süd)	active	Ötztal Alps, Austria	10.756°E, 46.894°N	LfCode 7RG1137 (Wagner et al., 2020a)	Krainer and Mostler (2002, 2006); Berger et al. (2004); Krainer et al. (2007); Hausmann et al. (2012); Klug et al. (2014); Rieder (2017); Heigert (2018a,b); Winkler et al. (2018b); Groh and Blöthe (2019); Wagner et al. (2019); Wagner et al. (2020)

(continued)

Table 5.2 (continued)

Name	State	Mountain range	Position	Rock Glacier Inventory ID	References
Inneres Bergli	active	Sammaun Group, Austria	10.251°E, 46.959°N	LfCode 7RG2811 (Wagner et al., 2020a)	Brodacz (2019); Winkler et al. (2018b); Wagner et al. (2019a, 2020)
Kaiserberg	active	Ötztal Alps, Austria	10.678°E, 46.910°N	LfCode 7RG1077 (Wagner et al., 2020a)	Krainer and Mostler (2002); Krainer et al. (2007); Hausmann et al. (2012); Klug et al. (2014); Brodacz (2019); Groh and Blöthe (2019)
Krummgampen	active	Ötztal Alps, Austria	10.700°E, 46.873°N	LfCode 7RG1055 (Wagner et al., 2020a)	Thies et al. (2013); Rogger et al. (2017); Wagner et al. (2019a; 2020b)
Langtal (syn. Hinteres Langtal-kar, Gößnitz)	active	Schober Group, Austria	12.781°E, 46.985°N	LfCode 2RG1280 (Wagner et al., 2020a)	Krainer and Mostler (2001, 2002); Krainer et al. (2007); Roer et al. (2008)
Las Argualas	active	Pyrenees, Spain	0.272°W, 42.774°N		Fabre et al. (1995)
Las Liebres	active	Chilean Andes, Chile	69.948°W, 30.251°S		Monnier and Kinnard (2015)
Lazaun	active	Ötztal Alps, Italy	10.754°E, 46.746°N	ID 904 (Autonome Provinz Bozen, 2018)	Krainer et al. (2015a)
Lost Lake	active	Sierra Blanca, Colorado	105.489°W, 37.608°N		Giardino et al. (1992)
Maya	inactive	Pennine Alps, Switzerland	7.496°E, 46.170°N		Tenthorey (1992, 1993 1994); Mari et al. (2013)

(continued)

Table 5.2 (continued)

Name	State	Mountain range	Position	Rock Glacier Inventory ID	References
Mount Śleża	relict	Śleża Massif, Poland	16.751°E, 50.837 N		Żurawek (2002)
Mount Tukuhnikivatz	active	La Sal Mountains, Utah	109.252°W, 38.446°N 109.259°W, 38.447°N		Geiger et al. (2014)
Muragl	active	Livigno Alps, Switzerland	9.930°E, 46.506°N		Vonder Mühll (1993); Arenson et al. (2002); Kääb et al. (2003); Musil et al. (2006); Springman et al. (2012); Cicoira et al. (2019)
Murfreit	active	Dolomites, Italy	11.808°E, 46.535°N	ID 994 (Autonome Provinz Bozen, 2018)	Krainer et al. (2012)
Murtèl-Corvatsch	active	Bernina Group, Switzerland	9.822°E, 46.428°N		King et al. (1987); Haeberli (1985); Haeberli et al. (1988); Vonder Mühll (1992, 1993); Speck (1994); Barsch (1996); Arenson et al. (2002, 2010); Kääb et al. (2003); Springman et al. (2012); Scherler et al. (2013, 2014); Cicoira et al. (2019);
Nair	active	Albula Alps, Switzerland	9.788°E, 46.509°N		Ikeda and Matsuoka (2006); Emmert and Kneisel (2017)
Pontresina-Schafberg (syn. Foura da l'amd Ursina)	active	Livigno Alps, Switzerland	9.928°E, 46.498°N		Vonder Mühll (1993); Arenson et al. (2002); Zenklusen Mutter and Phillips (2012); Kenner et al. (2016, 2020); Cicoira et al. (2019)

(continued)

Table 5.2 (continued)

Name	State	Mountain range	Position	Rock Glacier Inventory ID	References
Reichenkar	active	Stubai Alps, Austria	11.026°E, 47.044°N	LfCode 7RG1978 (Wagner et al., 2020a)	Krainer et al. (2000, 2007); Krainer and Mostler (2002); Hausmann et al. (2007, 2012); Klug et al. (2014); Brodacz (2019; Wagner et al. (2020)
Ritigraben	active	Pennine Alps, Switzerland	7.850 E, 46.175 N		Zenklusen Mutter and Phillips (2012); Cicoira et al. (2019); Kenner et al. (2020)
Sainte-Anne	active	Cottian Alps, France	6.806°E, 44.610°N		Evin (1983); Evin and Fabre (1990); Assier et al. (1996)
Schöneben	relict	Seckau Alps, Austria	14.673°E, 47.374°N	LfCode 6RG681 (Wagner et al., 2020a)	Untersweg and Schwendt (1995, 1996); Pauritsch (2011); Pauritsch et al. (2015, 2017); Winkler et al. (2016a,b); Brodacz (2019); Wagner et al. (2019a,b, 2020b)
Slims River	active	St. Elias Mountains, Yukon	138.576°W, 60.954°N		Blumstengel and Harris (1988); Harris et al. (1994)
Steintälli	active	Pennine Alps, Switzerland	7.830°E, 46.128 N		Wirz et al. (2016)
Stepanek	active	Cordón del Plata, Argentina	69.382°W, 32.969°S		Liaudat et al. (2020)
Suvretta	active	Albula Alps, Switzerland	9.783°E, 46.492°N		Vonder Mühll (1993); Barsch (1996)
Tsavolire	active	Pennine Alps, Switzerland	7.510°E, 46.166°N		Tenthorey (1992, 1993, 1994)
Uertsch	inactive	Albula Alps, Switzerland	9.844°E, 46.605°N		Emmert and Kneisel (2017)

Conclusion

6

This thesis integrates available data and new results into a consistent hydrogeological conceptual model of the rock glacier Innere Ölgrube. The model reflects site specific characteristics of the rock glacier and its catchment derived from a multi-disciplinary approach, provides a comprehensive basis for a prospective distributed parameter model at the rock glacier scale, and is constrained by a lumped parameter rainfall runoff model characterizing the catchment scale. It complies with the downward approach to hydrological prediction by attempting to explore the dominant processes identified to control the catchment response by Wagner et al. (in prep.) and characterize them at the rock glacier scale (Grayson and Blöschl, 2000; Sivapalan et al., 2003).

The internal structure of the rock glacier is depicted by a 3D geometrical model based on a combination of geophysical measurements, surface elevation and displacement rates, and simple creep modelling. Continuous permafrost prevails in the rock glacier rooting zone, while permafrost is degrading close to the rock glacier front. The contrasting hydraulic properties of adjacent frozen and unfrozen debris and the heterogeneous morphology of the permafrost table exert a strong influence on the groundwater flow field within the rock glacier. During the summer months the major hydrostratigraphic units constituting the rock glacier include the active layer, the permafrost layer, and the base layer. The active layer is highly permeable (hydraulic conductivity on the order of $\sim 10^{-2}$ m s^{-1}) and characterized by pronounced heterogeneity including preferential flow paths that represent the fast flow component. It exhibits a small but significant storage capacity that is mainly attributed to depressions and swells shaped by the permafrost table. The unfrozen base layer constitutes the major storage component of the rock glacier and heavily affects the catchment response by representing the rate-controlling element. It governs the slow flow component of the rock

S. Seelig, *Characterizing Groundwater Flow Dynamics and Storage Capacity in an Active Rock Glacier*, BestMasters, https://doi.org/10.1007/978-3-658-37073-2_6

137

glacier and exhibits a hydraulic conductivity on the order of $\sim 10^{-5}$ m s^{-1}. The permafrost layer separates the highly permeable perched aquifer provided by the active layer above from the large storage reservoir of the unfrozen base layer below. This feature is attributed to its comparatively low hydraulic conductivity, which is generally $> 10^{-8}$ m s^{-1} in the rooting zone where continuous permafrost conditions prevail, but presumably orders of magnitude higher close to the rock glacier front. It exhibits low storage capacity due to the presence of interstitial ice but allows for slow percolation of water contributing to the slow flow component. Preferential flow paths provided by crevasses, taliks, and thermokarst channels in the lower part of the rock glacier add to the fast flow component.

Distributed recharge occurs across the rock glacier surface and by groundwater flux from the surrounding talus and moraines. Localized infiltration of cirque glacier meltwater and catchment groundwater occurs in the northern part of the rock glacier rooting zone. Due to its large storage capacity and heterogeneous flow path geometry, mixing of water from different sources takes places at several locations within the rock glacier exhibiting distinct hydraulic properties.

The rock glacier Innere Ölgrube shares hydrogeological characteristics with many active rock glaciers. This suggests that some conclusions drawn from this specific site are likely to be representative for active rock glaciers in general, although this remains to be proven by future studies. Specifically, the recharge-dominated short term response of the rock glacier spring reflects catchment characteristics (in particular, the presence of a cirque glacier in the uppermost part of the catchment). The base layer hydraulic properties are similar to that of the relict Schöneben rock glacier, suggesting similar base flow characteristics of rock glacier regardless of the state of activity. This has important consequences for the predictability of future alpine catchment behavior in the light of climate change.

Outlook

7

The results presented in this thesis represent an intermediate step in the quantitative characterization of the Innere Ölgrube catchment and provide a basis for a prospective distributed parameter model complementing the lumped-parameter rainfall-runoff model provided by Wagner et al. (in prep.). The specified aquifer geometry and boundary conditions can be easily integrated into a numerical groundwater flow model. The estimated hydraulic properties provide prior information and initial estimates for a corresponding parameter estimation routine and uncertainty analysis. The taken approach and obtained results aim at maximal comparability to the numerical groundwater flow model of the relict Schöneben rock glacier given by Pauritsch et al. (2017) in order to facilitate comparative analysis of active and relict rock glacier hydraulics, respectively.

However, challenges regarding the implementation of such a model remain to be met. Little information is available regarding the drainable porosity of the major aquifers constituting the rock glacier. Improved estimates might be obtained from recession analysis based on solutions of the groundwater flow equation (e. g. Kovács et al., 2005; Hergarten and Birk, 2007; Birk and Hergarten, 2010; Pauritsch et al., 2015; Winkler et al., 2016a). If combined with the analysis presented in this thesis, the estimates derived here might be checked and complemented.

Note that all four layers depicted in fig. 4.1 exhibit steep zones. These might induce a serious challenge to a prospective groundwater flow model in case a vertically distorted finite difference or control volume finite difference (CVFD) grid is employed (i. e. each cell within a layer is assigned different top and bottom elevations and, consequently, layer thickness varies from cell to cell; Anderson et al., 2015; Langevin et al., 2017). Such a grid design ensures that model layers represent spatial domains exhibiting relatively uniform hydraulic properties and

S. Seelig, *Characterizing Groundwater Flow Dynamics and Storage Capacity in an Active Rock Glacier*, BestMasters,
https://doi.org/10.1007/978-3-658-37073-2_7

small vertical losses in hydraulic head, thereby increasing accuracy of the model calculations (since the discretized form of the governing equations is based on these assumptions, while errors introduced by the vertical distortion are typically small for groundwater flow models; Anderson et al., 2015; Langevin et al., 2017). However, in order to achieve accurate solutions, cell connections are subject to geometric constraints (Narasimhan and Witherspoon, 1976; Anderson et al., 2015; Langevin et al., 2017). A line connecting adjacent nodes (cell centers) is required to intersect the shared face between these cells at a right angle as well as at an appropriate mean position of this face. As outlined by Langevin et al. (2017) it is the user's responsibility to ensure that these requirements are met by the specified grid. Large deviations from these requirements introduce large errors in simulated flows and heads (Panday et al., 2013; Langevin et al., 2017). These errors generally decrease as resolution increases, but are difficult to quantify because they are not apparent from the simulated water budget (Panday et al., 2013). Large vertical offsets between horizontally adjacent cells may violate these geometric requirements (Langevin et al., 2017). The regular CVFD grid implies that the principal components of the hydraulic conductivity tensor are aligned parallel to the model coordinate system. In steeply dipping layers (> 10°), however, the orientation of the hydraulic connection between adjacent cells may seriously deviate from the principal axes of the model coordinate system, introducing a significant error in model calculations (Anderson et al., 2015). These complications are likely to affect a prospective groundwater flow model in the steep zones outlined in fig. 4.1. Potential approaches to solve this problem include the Ghost Node Correction Package and the XT3D option, which are both available for MODFLOW 6 (Langevin et al., 2017; Provost et al., 2017).

If included, representation of preferential flow paths is potentially challenging: The suspected nonlinear flow features characterizing some parts of the rock glacier groundwater flow field might be modeled using the Connected Linear Network Package currently under development for MODFLOW 6, but already available from the MODFLOW development team upon request (R. A. Collenteur, pers. comm.). However, the currently available data may not allow for such a level of complexity, and accordingly serious problems regarding convergence and calibration issues might be expected. Alternatively, heterogeneous layers might be specified along the lines of thought outlined by Pauritsch et al. (2017), making use of unstructured discretization options if necessary (e. g. Discretization by Vertices Package, Unstructured Discretization Package, both available for MODFLOW 6). Likely, data scarcity and manageable complexity need to be balanced against the heterogeneous nature of the active rock glacier.

Due to the limited data available in the alpine catchment, equifinality is likely to play a key role in model calibration and uncertainty analysis. Model uniqueness might be increased by coupling a transport model to a prospective groundwater flow model, using natural and artificial tracer data as additional calibration target (increasing the diversity of available targets; cf. Hunt et al., 2006; Hill and Tiedeman, 2007; Rehrl and Birk, 2010; Seelig, in prep.).

While cirque glacier melt water dominates the total ice melt flux within the catchment, the contribution by melting permafrost ice remains poorly constrained (besides being small compared to glacial meltwater). An update of the boundary conditions derived in this thesis might be achieved by including vertical surface displacement rates as indicator of changing subsurface ice volumes as demonstrated by Kääb et al. (2003), Klug et al. (2014), and Groh and Blöthe (2019). Specifically, Groh and Blöthe (2019) demonstrated the applicability of an image tracking approach to determine degrading permafrost at the nearby Kaiserberg Rock Glacier. Since the creep model employed in this thesis to derive the internal rock glacier structure is based on surface displacement rates determined by Groh and Blöthe (2019) using the same image tracking approach, adapting their procedure would promote a high degree of methodological consistency.

The geometrical model provided in this thesis is subject to several uncertainties and might be refined in a variety of ways. The spatially heterogeneous ice content evaluated by Hausmann et al. (2012) might be included in the constitutive equation (eq. 3.6) of the employed creep model (cf. Arenson and Springman, 2005). The performance might be seriously improved by including extending and compressive deformation regimes in the creep model (as is common in glacier mechanics; c. f. Cuffey and Paterson, 2010). Alternatively, more sophisticated models might be employed that are able to capture the complex deformation characteristics of the rock glacier Innere Ölgrube at a more appropriate level of accuracy (cf. Haeberli et al., 2006; Jensen and Hergarten, 2006; Springman et al., 2012). In addition, the heterogeneous internal structure of the permafrost layer might be explored by combining high resolution refraction seismics and electrical resistivity measurements as demonstrated by Hauck et al. (2011) and Mewes et al. (2017).

Several authors have stressed the similarity between karst systems and rock glaciers (Winkler et al., 2016a,b; Pauritsch et al., 2017; Heigert, 2018; Wagner et al., 2020b), and a range of methods well established in karst hydrology have been adapted to the investigation of rock glaciers (e. g. Winkler et al., 2016a,b; Heigert, 2018; Wagner et al., 2020b; this thesis). Besides intrinsic similarities (e. g. strong heterogeneity), karst systems and alpine crystalline catchments share methodological restrictions (e. g. scarce data availability; cf. Rehrl and Birk,

2010), potentially enhancing the compability of methods. Potential future applications might include the modeling of turbulent flow (e. g. Shoemaker et al., 2008; Reimann et al., 2011; Mayaud et al., 2015) and the development and impact of preferential flow path networks (e. g. Bauer et al., 2003; Birk et al., 2003, 2005, 2006; Liedl et al., 2003).

The presented conceptualization of thermokarst lake drainage is probably an oversimplification. Small changes in boundary conditions (e. g. climate) suffice to initiate thermokarst processes, enabling the development of drainage channels (Kääb and Haeberli, 2001; Vonder Mühll et al., 2003). Thermokarst features are typically developed along the depressions and furrows characterizing rock glacier surfaces due to downgradient flow (and associated convective heat transport) (Buchli et al., 2013). Several conspicuous features of the water level record are potentially related to such processes and thus require further analysis. For example, fig. 3.13a reveals two distinct water level drops around 10.8.2015 and 15.8.2015, respectively, which are not explainable by the simple interpretation given in this thesis. Note also the slight diurnal variations observable during some of the evaluation periods (fig. 3.13). These might actually indicate some sort of melting process affecting the lake water balance during these periods, thereby violating the assumptions underlying the applied simple calculation procedure. As a rather vague working hypothesis these processes in summary support the conclusion that the results given in tab. 4.2 are lower bound estimates of the actual hydraulic conductivity. However, a more sophisticated analysis of these processes is desirable.

Combined application of a lumped-parameter rainfall-runoff model (GR4J+) at the catchment scale and a prospective distributed parameter model (e. g. MODFLOW) at the rock glacier scale conforms to the downward approach to hydrological prediction (Sivapalan et al., 2003). Accordingly, once these prove successful (the former already did so, the latter remains to be proven), complexity might be increased stepwise (e. g. by more sophisticated modeling of snow hydrology (e. g. UEB, SNOWGRID, AMUNDSEN) and/or glacier hydrology (e. g. Arnold, 1996); cf. Hood and Hayashi, 2015). Additional rock glacier dominated catchments might be analyzed in the same manner in order to establish a general procedure for alpine catchment characterization based on these approaches.

On a regional scale, the ice content (water equivalent) stored in rock glaciers might be assessed by combining image tracking (e. g. Groh and Blöthe, 2019) with the creep model presented above (or any improved version thereof) which requires knowledge of the surface displacement rate, a digital surface elevation model, and a reasonable porosity/bulk density estimate. The clearly defined physical processes and parameters support the credibility of such an assessment (Jones et al., 2019) and allow for a systematic improvement of the model itself (c. f. Arenson and Springman, 2005).

Diurnal variations in discharge and natural tracer time series recorded at rock glacier springs during late summer indicate substantial recharge from cirque glaciers present in the higher parts of the catchment, as demonstrated qualitatively by Potter (1972), Tenthorey (1993), Krainer and Mostler (2002) and Wagner et al. (2019a). Spectral analysis provides a tool to test this hypothesis quantitatively by comparing individual rock glaciers (including or lacking cirque glaciers in their catchments) at a specified significance level, using the periodograms as simple statistical tests (Scargle, 1982; Birk et al., 2004). The robust method employed in this thesis tolerates data gaps and irregular sampling intervals, is applicable to any record if physically meaningful, and has proven earlier to allow for an identification of localized recharge components (Birk et al., 2004). Such an approach might be extended by comparing time series analyses, rainfall-runoff modelling (e. g. GR4J+), and melting rates of rock glacier permafrost ice obtainable from vertical surface displacement rates (e. g. derived from image tracking) to test the hypothesis that cirque glacier recharge outweighs recharge from melting permafrost ice. Results might be integrated in the multi component mixing model for rock glacier recharge provided by Winkler et al. (2018b).

Correction to: Characterizing Groundwater Flow Dynamics and Storage Capacity in an Active Rock Glacier

Correction to:
S. Seelig, *Characterizing Groundwater Flow Dynamics and Storage Capacity in an Active Rock Glacier*, BestMasters, https://doi.org/10.1007/978-3-658-37073-2

The Author has changed his name from Simon Kainz to Simon Seelig, The original version of this book has been revised.

The updated version of the book can be found at
https://doi.org/10.1007/978-3-658-37073-2

References

Andersland, O. B.; Wiggert, D. C.; Davies, S. H. (1996): Hydraulic Conductivity of Frozen Granular Soils. Journal of Environmental Engineering 122 (3), 212–216.

Anderson, M. P.; Woessner, W. W.; Hunt, R. J. (2015): Applied Groundwater Modeling. Simulation of Flow and Advective Transport. 2nd ed.. Elsevier, Amsterdam.

Arenson, L. U. (2002): Unstable Alpine Permafrost: A Potentially Important Natural Hazard. Variations of Geotechnical Behaviour with Time and Temperature. Doctoral Thesis. Eidgenössische Technische Hochschule, Zürich.

Arenson, L.; Hoelzle, M.; Springman, S. (2002): Borehole deformation measurements and internal structure of some rock glaciers in Switzerland. Permafrost Periglac. Process. 13 (2), 117–135.

Arenson, L. U.; Johansen, M. M.; Springman, S. M. (2004): Effects of volumetric ice content and strain rate on shear strength under triaxial conditions for frozen soil samples. Permafrost Periglac. Process. 15 (3), 261–271.

Arenson, L. U.; Springman, S. M. (2005): Mathematical descriptions for the behaviour of ice-rich frozen soils at temperatures close to 0 °C. Can. Geotech. J. 42 (2), 431–442.

Arenson, L. U.; Jakob, M. (2010): The significance of rock glaciers in the dry Andes—A discussion of Azócar and Brenning (2010) and Brenning and Azócar (2010). Permafrost Periglac. Process. 21 (3), 282–285.

Arenson, L. U.; Hauck, C.; Hilbich, C.; Seward, L.; Yamato, Y.; Springman, S. (2010): Subsurface Heterogeneities in the Murtèl – Corvatsch Rock Glacier, Switzerland. In: Canadian Geotechnical Society (Ed.): Proceedings of the joint 63rd Canadian Geotechnical Conference and the 6th Canadian Permafrost Conference. Calgary, Alberta, 1494–1500.

Arnold, N.; Richards, K.; Willis, I.; Sharp, M. (1998): Initial results from a distributed, physically based model of glacier hydrology. Hydrol. Process. 12 (2), 191–219.

Assier, A.; Fabre, D.; Evin, M. (1996): Prospection Électrique sur les Glaciers Rocheux du Cirque de Sainte-Anne (Queyras, Alpes du Sud, France). Permafrost Periglac. Process. 7, 53–67.

Atkinson, T. C.; Smith, D. I.; Lavis, J. J.; Whitaker, R. J. (1973): Experiments in Tracing Underground Waters in Limestones. Journal of Hydrology 19, 323–349.

Ban, N.; Schmidli, J.; Schär, C. (2015): Heavy precipitation in a changing climate. Does short-term summer precipitation increase faster? Geophys. Res. Lett. 42 (4), 1165–1172.

© The Editor(s) (if applicable) and The Author(s), under exclusive license to Springer Fachmedien Wiesbaden GmbH, part of Springer Nature 2022
S. Seelig, *Characterizing Groundwater Flow Dynamics and Storage Capacity in an Active Rock Glacier*, BestMasters,
https://doi.org/10.1007/978-3-658-37073-2

Barnett, T. P.; Adam, J. C.; Lettenmaier, D. P. (2005): Potential impacts of a warming climate on water availability in snow-dominated regions. Nature 438 (7066), 303–309.

Barsch, D.; Fierz, H.; Haeberli, W. (1979): Shallow Core Drilling and Bore-Hole Measurements in the Permafrost of an Active Rock Glacier near the Grubengletscher, Wallis, Swiss Alps. Arctic and Alpine Research 11 (2), 215–228.

Barsch, D. (1996): Rockglaciers. Indicators for the Present and Former Geoecology in High Mountain Environments. Springer, Berlin, Heidelberg.

Bauer, S.; Liedl, R.; Sauter, M. (2003): Modeling of karst aquifer genesis: Influence of exchange flow. Water Resources Research 39 (10).

Bear, J. (1972): Dynamics of fluids in porous media. Elsevier, New York.

Bear, J.; Bachmat, Y. (1990): Introduction to Modeling of Transport Phenomena in Porous Media. Kluwer Academic Publishers, Dordrecht (Theory and Applications of Transport in Porous Media, 4).

Bear, J.; Cheng, A. H.-D. (2010): Modeling Groundwater Flow and Contaminant Transport. Springer, Dordrecht, Heidelberg, London, New York (Theory and Applications of Transport in Porous Media, 23).

Beniston, M.; Farinotti, D.; Stoffel, M.; Andreassen, L. M.; Coppola, E.; Eckert, N.; Fantini, A.; Giacona, F.; Hauck, C.; Huss, M.; Huwald, H.; Lehning, M.; López-Moreno, J.-I.; Magnusson, J.; Marty, C.; Morán-Tejéda, E.; Morin, S.; Naaim, M.; Provenzale, A.; Rabatel, A.; Six, D.; Stötter, J.; Strasser, U.; Terzago, S.; Vincent, C. (2018): The European mountain cryosphere. A review of its current state, trends, and future challenges. The Cryosphere 12 (2), 759–794.

Berger, J.; Krainer, K.; Mostler, W. (2004): Dynamics of an active rock glacier (Ötztal Alps, Austria). Quaternary Research 62 (3), 233–242.

Bevilacqua, I. (2019): Thermal Regime of the Uppermost Blocky Layer of Intact and Relict Rock Glaciers. Master Thesis. Karl-Franzens-Universität, Graz.

Birk, S. (2002): Characterisation of Karst Systems by Simulating Aquifer Genesis and Spring Responses: Model Development and Application to Gypsum Karst. Eberhard Karls Universität Tübingen, Tübingen (Tübinger Geowissenschaftliche Arbeiten (TGA), Reihe C).

Birk, S.; Liedl, R.; Sauter, M.; Teutsch, G. (2003): Hydraulic boundary conditions as a controlling factor in karst genesis. A numerical modeling study on artesian conduit development in gypsum. Water Resour. Res. 39 (1).

Birk, S.; Liedl, R.; Sauter, M. (2004): Identification of localised recharge and conduit flow by combined analysis of hydraulic and physico-chemical spring responses (Urenbrunnen, SW-Germany). Journal of Hydrology 286 (1-4), 179–193.

Birk, S.; Geyer, T.; Liedl, R.; Sauter, M. (2005): Process-Based Interpretation of Tracer Tests in Carbonate Aquifers. Ground Water 43 (3), 381–388.

Birk, S.; Liedl, R.; Sauter, M. (2006): Karst spring responses examined by process-based modeling. Ground Water 44 (6), 832–836.

Birk, S.; Hergarten, S. (2010): Early recession behaviour of spring hydrographs. Journal of Hydrology 387 (1-2), 24–32.

Blumstengel, W. K.; Harris, S. A. (1988): Observations on an active lobate rock glacier, Slims River valley, St. Elias Range, Canada. In: Kaare, S. (ed.): Proceedings of the 5th International Conference on Permafrost, Trondheim, Norway. Tapir Publ., Trondheim, 689–694.

Boeckli, L.; Brenning, A.; Gruber, S.; Noetzli, J. (2012a): Permafrost distribution in the European Alps. Calculation and evaluation of an index map and summary statistics. The Cryosphere 6 (4), 807–820.

Boeckli, L.; Brenning, A.; Gruber, A.; Noetzli, J. (2012b): Alpine permafrost index map. PANGAEA, https://doi.org/10.1594/PANGAEA.784450, last accessed 20.11.2020.

Box, G. E. P.; Jenkins, G. M.; Reinsel, G. C. (2016): Time series analysis. Forecasting and control. 5th ed. (Wiley series in probability and statistics).

Brodacz, A. (2019): Speicherverhalten und Entwässerungsdynamik von Blockgletschern – Vergleichende Untersuchungen zwischen intakten und reliktischen Blockgletschern. Master Thesis. Karl-Franzens-Universität, Graz.

Brown, W. H. (1925): A probable fossil rock glacier. Journal of Geology, 33, 464–466.

Brutsaert, W. (1994): The unit response of groundwater outflow from a hillslope. Water Resour. Res. 30 (10), 2759–2763.

Buchli, T.; Merz, K.; Zhou, X.; Kinzelbach, W.; Springman, S. M. (2013): Characterization and Monitoring of the Furggwanghorn Rock Glacier, Turtmann Valley, Switzerland. Results from 2010 to 2012. Vadose Zone Journal 12 (1).

Buchli, T.; Kos, A.; Limpach, P.; Merz, K.; Zhou, X.; Springman, S. M. (2018): Kinematic investigations on the Furggwanghorn Rock Glacier, Switzerland. Permafrost Periglac. Process. 29 (1), 3–20.

Buckel, J.; Otto, J.-C. (2018): The Austrian Glacier Inventory GI 4 (2015) in ArcGis (shapefile) format. PANGAEA, https://doi.org/10.1594/PANGAEA.887415, last accessed 20.11.2020.

Burger, K. C.; Degenhardt, J. J.; Giardino, J. R. (1999): Engineering geomorphology of rock glaciers. Geomorphology 31 (1-4), 93–132.

Burt, T. P.; Williams, P. J. (1976): Hydraulic conductivity in frozen soils. Earth Surf. Process. 1, 349–360.

Busch, K.-F.; Luckner, L.; Tiemer, K.; Mattheß, G. (1993): Geohydraulik. 3rd ed.. Borntraeger, Berlin (Lehrbuch der Hydrogeologie, 3).

Chatwin, P. C. (1971): On the interpretation of some longitudinal dispersion experiments. J. Fluid Mech. 48 (4), 689–702.

Cicoira, A.; Beutel, J.; Faillettaz, J.; Gärtner-Roer, I.; Vieli, A. (2019): Resolving the influence of temperature forcing through heat conduction on rock glacier dynamics. A numerical modelling approach. The Cryosphere 13 (3), 927–942.

Clark, I.; Fritz, P. (1997): Environmental Isotopes in Hydrogeology. Lewis Publishers, Boca Raton, New York.

Clow, D. W.; Schrott, L.; Webb, R.; Campbell, D. H.; Torizzo, A.; Dornblaser, M. (2003): Ground Water Occurrence and Contributions to Streamflow in an Alpine Catchment, Colorado Front Range. Ground Water 41 (7), 937–950.

Colgan, W.; Rajaram, H.; Abdalati, W.; McCutchan, C.; Mottram, R.; Moussavi, M. S.; Grigsby, S. (2016): Glacier crevasses. Observations, models, and mass balance implications. Rev. Geophys. 54 (1), 119–161.

Colombo, N.; Gruber, S.; Martin, M.; Malandrino, M.; Magnani, A.; Godone, D.; Freppaz, M.; Fratianni, S.; Salerno, F. (2018a): Rainfall as primary driver of discharge and solute export from rock glaciers. The Col d'Olen Rock Glacier in the NW Italian Alps. Science of The Total Environment 639, 316–330. DOI: https://doi.org/10.1016/j.scitotenv.2018. 05.098.

Colombo, N.; Salerno, F.; Gruber, S.; Freppaz, M.; Williams, M.; Fratianni, S.; Giardino, M. (2018b): Review. Impacts of permafrost degradation on inorganic chemistry of surface fresh water. Global and Planetary Change 162, 69–83.

Colombo, N.; Sambuelli, L.; Comina, C.; Colombero, C.; Giardino, M.; Gruber, S.; Viviano, G.; Vittori Antisari, L.; Salerno, F. (2018c): Mechanisms linking active rock glaciers and impounded surface water formation in high-mountain areas. Earth Surface Processes and Landforms 43 (2), 417–431.

Copernicus Land Monitoring Service (2015): European Digital Elevation Model (EU-DEM) v1.1. http://land.copernicus.eu/pan-european/satellite-derived-products/eu-dem/eu-dem, last accessed 20.11.2020

Corte, A. E. (1987): Central Andes rock glaciers: applied aspects. In: Giardino, J.R., Shroder, J.F. Jr., Vitek, J.D. Eds.., Rock Glaciers. Allen and Unwin, London, 289–304.

Covington, M. D.; Wicks, C. M.; Saar, M. O. (2009): A dimensionless number describing the effects of recharge and geometry on discharge from simple karstic aquifers. Water Resour. Res. 45 (11), 161.

Covington, M. D.; Banwell, A. F.; Gulley, J.; Saar, M. O.; Willis, I.; Wicks, C. M. (2012): Quantifying the effects of glacier conduit geometry and recharge on proglacial hydrograph form. Journal of Hydrology 414-415, 59–71.

Croce, F. A.; Milana, J. P. (2002): Internal structure and behaviour of a rock glacier in the Arid Andes of Argentina. Permafrost Periglac. Process. 13 (4), 289–299.

Cuffey, K. M.; Paterson, W. S. B. (2010): The Physics of Glaciers. 4th ed.. Elsevier Butterworth Heinemann, Burlington, Mass.

Darcy, H. (1856): Les fontaines publiques de la ville de Dijon. Exposition et application des principes a suivre et des formules a employer dans les questions de distribution d'eau; ouvrage terminé par un appendice relatif aux fournitures d'eau de plusieurs villes au filtrage des eaux à la fabrication des tuyaux de fonte, de plomb, de tole et de bitume. Dalmont, Paris.

Davis, J. C. (2002): Statistical and Data Analysis in Geology. 3rd ed.. Wiley.

Davis, P. M.; Atkinson, T. C.; Wigley, T. M. L. (2000): Longitudinal disperion in natural channels. 2. The roles of shear flow dispersion and dead zones in the River Severn, U.K. Hydrology and Earth System Sciences 4 (3), 355–371.

Day, T. J. (1975): Longitudinal Dispersion in Natural Channels. Water Resour. Res. 11 (6), 909–918.

Drucker, D. C. (1967): Introduction to Mechanics of Deformable Solids. McGraw-Hill Book Company, New York.

Emmert, A.; Kneisel, C. (2017): Internal structure of two alpine rock glaciers investigated by quasi-3-D electrical resistivity imaging. The Cryosphere 11 (2), 841–855.

Environmental Systems Research Institute (ESRI) (2016): ArcGIS Desktop 10.4. Redlands, CA.

Evin, M.; Assier, A. (1981): Glacier et glaciers rocheux dans le Haut-Vallon du Loup (Haute-Ubaye, Alpes du Sud, France). Zeitschrift für Gletscherkunde und Glaziologie, 19, 27-41.

Evin, M.; Fabre, D. (1990): The Distribution of Permafrost in Rock Glaciers of the Southern Alps (France). Geomorphology 3, 57–71.

Fabre, D.; Garcia, F.; Evin, M.; Martinez, R.; Serrano, E.; Assier, A.; Smiraglia, C. (1995): Structure interne du glacier rocheux actif de Las Arguales (Pyrénées aragonaises, Espagne). La Houille Blanche (5–6), 144–147.

Field, M. S.; Pinsky, P. F. (2000): A two-region nonequilibrium model for solute transport in solution conduits in karstic aquifers. Journal of Contaminant Hydrology 44 (3–4), 329–351.

Field, M. S. (2002): The QTRACER2 Program for Tracer-Breakthrough Curve Analysis for Tracer Tests in Karstic Aquifers and Other Hydrologic Systems. United States Environmental Protection Agency (EPA), Washington, D.C. (EPA, 600/R-02/001).

Fisch, W., Sen.; Fisch, W., Jr.; Haeberli, W. (1977): Electrical D.C. resistivity soundings with long profiles on rock glaciers and moraines in the Alps of Switzerland. Zeitschrift für Gletscherkunde und Glazialgeologie 13 (1), 239–260.

Geiger, S. T.; Daniels, J. M.; Miller, S. N.; Nicholas, J. W. (2014): Influence of Rock Glaciers on Stream Hydrology in the La Sal Mountains, Utah. Arctic, Antarctic, and Alpine Research 46 (3), 645–658.

Geyer, T.; Birk, S.; Licha, T.; Liedl, R.; Sauter, M. (2007): Multitracer test approach to characterize reactive transport in karst aquifers. Ground Water 45 (1), 36–45.

Geyer, T. (2008): Process-based characterisation of flow and transport in karst aquifers at catchment scale. Doctoral Thesis. Georg-August-Universität, Göttingen.

Giardino, J. R.; Vitek, J. D. (1988): Interpreting the Internal Fabric of a Rock Glacier. Geografiska Annaler. Series A, Physical Geography 70 (1), 15–25.

Giardino, J. R.; Vitek, J. D.; DeMorett, J. L. (1992): A Model of Water Movement in Rock Glaciers and Associated Water Characteristics. In: Dixon, J. C.; Abrahams, A. D. (ed.): Periglacial Geomorphology. Wiley, Chichester, 159–184.

Glen, J. W. (1955): The creep of polycrystalline ice. Proc. R. Soc. Lond. A 228 (1175), 519–538.

Glen, J. W. (1958): The flow law of ice. A discussion of the assumptions made in glacier theory, their experimental foundations and consequences. Int. Assoc. Hydrol. Sci. Publ. 47, 171–183.

Gobiet, A.; Kotlarski, S.; Beniston, M.; Heinrich, G.; Rajczak, J.; Stoffel, M. (2014): 21st century climate change in the European Alps—a review. The Science of the total environment 493, 1138–1151.

Goltz, M. N.; Roberts, P. V. (1986): Interpreting organic solute transport data from a field experiment using physical nonequilibrium models. Journal of Contaminant Hydrology 1 (1-2), 77–93.

Graßmair, R.; Erschbamer, B. (2015): Die Besiedelung des Blockgletschers Äußeres Hochebenkar im Vergleich zur angrenzenden Vegetation. In: Schallhart, N.; Erschbamer, B. (ed.): Forschung am Blockgletscher. Methoden und Ergebnisse. Innsbruck Univ. Press, Innsbruck (Alpine Forschungsstelle Obergurgl, 4), 159–180.

Groh, T.; Blöthe, J. H. (2019): Rock Glacier Kinematics in the Kaunertal, Ötztal Alps, Austria. Geosciences 9 (9), 373.

Haeberli, W.; Patzelt, G. (1982): Permafrostkartierung im Gebiet des Hochebenkar-Blockgletscher, Obergurgl, Ötztaler Alpen. Zeitschrift für Gletscherkunde und Glazialgeologie 18 (2), 127–150.

Haeberli, W. (1985): Creep of mountain permafrost. Internal structure and flow of alpine rock glaciers. Eidgenössische Technische Hochschule, Zürich (Mitt. der Versuchsanstalt für Wasserbau, Hydrologie und Glaziologie der ETH Zürich, 77).

Haeberli, W.; Huder, J.; Keusen, H.-R.; Pika, J.; Röthlisberger, H. (1988): Core drilling through rock glacier-permafrost. In: Kaare, S. (ed.): Proceedings of the 5th International Conference on Permafrost, Trondheim, Norway. Tapir Publ., Trondheim, 937–942.

Haeberli, W.; Beniston, M. (1998): Climate Change and Its Impacts on Glaciers and Permafrost in the Alps. AMBIO A Journal of the Human Environment 27 (4), 258–265.

Haeberli, W.; Hoelzle, M.; Kääb, A.; Keller, F.; Vonder Mühll, D.; Wagner, S. (1998): Ten years after drilling through the permafrost of the active rock glacier Murtèl, Eastern Swiss Alps: Answered questions and new perspectives. In: Lewkowicz, A. G.; Allard, M.; International Permafrost Association (ed.): Proceedings of the 8th International Conference on Permafrost. June 23–27, 1998, Yellowknife, Canada, 403–410.

Haeberli, W.; Kääb, A.; Vonder Mühll, D.; Teysseire, P. (2001): Prevention of outburst floods from periglacial lakes at Grubengletscher, Valais, Swiss Alps. Journal of Glaciology 47 (156), 111–122.

Haeberli, W.; Hallet, B.; Arenson, L.; Elconin, R.; Humlum, O.; Kääb, A.; Kaufmann, V.; Ladanyi, B.; Matsuoka, N.; Springman, S.; Mühll, D. V. (2006): Permafrost creep and rock glacier dynamics. Permafrost Periglac. Process. 17 (3), 189–214.

Hager, B.; Foelsche, U. (2015): Stable isotope composition of precipitation in Austria. AJES 108 (2).

Hambrey, M. J.; Lawson, W. (2000): Structural styles and deformation fields in glaciers: a review. In: Maltman, A. J.; Hubbard, B.; Hambrey, M. J. (ed.): Deformation of Glacial Materials. Geological Society of London (Special Publications, 176), 59–83.

Harrington, J. S.; Mozil, A.; Hayashi, M.; Bentley, L. R. (2018): Groundwater flow and storage processes in an inactive rock glacier. Hydrol. Process. 32 (20), 3070–3088.

Harrington, J. S. (2017): The Hydrogeology of a Rock Glacier and Its Effect on Stream Temperature. Master Thesis. University of Calgary, Calgary, Alberta.

Harris, C.; Vonder Mühll, D.; Isaksen, K.; Haeberli, W.; Sollid, J. L.; King, L.; Holmlund, P.; Dramis, F.; Guglielmin, M.; Palacios, D. (2003): Warming permafrost in European mountains. Global and Planetary Change 39 (3-4), 215–225.

Harris, S. A.; Blumstengel, W. K.; Cook, D.; Krouse, H. R.; Whitley, G. (1994): Comparison of the Water Drainage from an Active Near-Slope Rock Glacier and a Glacier, St. Elias Mountains, Yukon Territory. Erdkunde 48 (2), 81–91.

Hartl, L.; Fischer, A. (2015): Meteorologische Bedingungen und Strahlungsverhältnisse am Blockgletscher Äußeres Hochebenkar. In: Schallhart, N.; Erschbamer, B. (ed.): Forschung am Blockgletscher. Methoden und Ergebnisse. Innsbruck Univ. Press, Innsbruck (Alpine Forschungsstelle Obergurgl, 4), 97–115.

Hartl, L.; Fischer, A.; Klug, C.; Nicholson, L. (2016): Can a simple Numerical Model Help to Fine-Tune the Analysis of Ground-Penetrating Radar Data? Hochebenkar Rock Glacier as a Case Study. Arctic, Antarctic, and Alpine Research 48 (2), 377–393.

Hauck, C.; Böttcher, M.; Maurer, H. (2011): A new model for estimating subsurface ice content based on combined electrical and seismic data sets. The Cryosphere 5 (2), 453–468.

Hauns, M.; Jeannin, P.-Y.; Atteia, O. (2001): Dispersion, retardation and scale effect in tracer breakthrough curves in karst conduits. Journal of Hydrology 241 (3-4), 177–193.

Hausmann, H.; Krainer, K.; Brückl, E.; Mostler, W. (2007): Internal structure and ice content of Reichenkar rock glacier (Stubai Alps, Austria) assessed by geophysical investigations. Permafrost Periglac. Process. 18 (4), 351–367.

Hausmann, H.; Krainer, K.; Brückl, E.; Ullrich, C. (2012): Internal structure, ice content and dynamics of Ölgrube and Kaiserberg rock glaciers (Ötztal Alps, Austria) determined from geophysical surveys. AJES 105 (2), 12–31.

Hayashi, M. (2020): Alpine Hydrogeology. The Critical Role of Groundwater in Sourcing the Headwaters of the World. Ground Water 58 (4), 498–510.

Hedderich, J.; Sachs, L. (2018): Angewandte Statistik. Methodensammlung mit R. 16th ed.. Springer Spektrum, Berlin, Germany.

Heigert, K. (2018): Speicherverhalten und Abflussdynamik aktiver Blockgletscher am Beispiel Ölgrube Süd, Kaunertal. Master Thesis. Karl-Franzens-Universität, Graz.

Hergarten, S.; Birk, S. (2007): A fractal approach to the recession of spring hydrographs. Geophys. Res. Lett. 34 (11), 527.

Hill, M. C.; Tiedeman, C. R. (2007): Effective Groundwater Model Calibration. With Analysis of Data, Sensitivities, Predictions, and Uncertainty. John Wiley & Sons, Hoboken, New Jersey.

Hoinkes, G.; Thöni, M. (1993): Evolution of the Ötztal-Stubai, Scarl-Campo and Ulten Basement Units. In: Raumer, J. F.; Neubauer, F. (ed.): Pre-Mesozoic Geology in the Alps. Springer Berlin Heidelberg, 485–494.

Hood, J. L.; Hayashi, M. (2015): Characterization of snowmelt flux and groundwater storage in an alpine headwater basin. Journal of Hydrology 521, 482–497.

Ikeda, A.; Matsuoka, N. (2006): Pebbly versus bouldery rock glaciers. Morphology, structure and processes. Geomorphology 73 (3–4), 279–296.

Ikeda, A.; Matsuoka, N.; Kääb, A. (2008): Fast deformation of perennially frozen debris in a warm rock glacier in the Swiss Alps. An effect of liquid water. J. Geophys. Res. 113 (F1), 212.

Intergovernmental Panel on Climate Change (IPCC) (2019): IPCC Special Report on the Ocean and Cryosphere in a Changing Climate. Pörtner, H.-O.; Roberts, D. C.; Masson-Delmotte, V.; Zhai, P.; Tignor, M.; Poloczanska, E.; Mintenbeck, K.; Alegría, A.; Nicolai, M.; Okem, A.; Petzold, J.; Rama, B.; Weyer, N. M. (ed.).

Jansen, F.; Hergarten, S. (2006): Rock glacier dynamics. Stick-slip motion coupled to hydrology. Geophys. Res. Lett. 33 (10).

Jeannin, P.-Y.; Maréchal, J.-C. (1998): Dispersion and tailing of tracer plumes in a karstic system (Milandre, JU, Switzerland). In: Jeannin, P.-Y.: Structure et comportement hydraulique des aquifers karstiques. Dissertation, Université de Neuchâtel.

Joanes, D. N.; Gill, C. A. (1998): Comparing Measures of Sample Skewness and Kurtosis. The Statistician 47, 183–189.

Johnson, P. G. (1978): Rock glacier types and their drainage systems, Grizzly Creek, Yukon Territory. Can. J. Earth Sci. 15 (9), 1496–1507.

Jones, D. B.; Harrison, S.; Anderson, K.; Whalley, W. B. (2019): Rock glaciers and mountain hydrology. A review. Earth-Sci. Rev. 193, 66–90.

Kääb, A.; Haeberli, W.; Gudmundsson, H. (1997): Analysing the Creep of Mountain Permafrost using High Precision Aerial Photogrammetry: 25 Years of Monitoring Gruben Rock Glacier, Swiss Alps. Permafrost Periglac. Process. 8, 409–426.

Kääb, A.; Haeberli, W. (2001): Evolution of a High-Mountain Thermokarst Lake in the Swiss Alps. Arctic, Antarctic, and Alpine Research 33 (4), 385–390.

Kääb, A.; Kaufmann, V.; Ladstädter, R.; Eiken, T. (2003): Rock glacier dynamics: implications from high-resolution measurements of surface velocity fields. In: Phillips, M.;

Springman, S. M.; Arenson, L. U. (ed.): Proceedings of the 8th International Conference on Permafrost. Vol. 2. Balkema, Lisse, 501–506.

Käss, W. (2004): Geohydrologische Markierungstechnik. 2nd ed.. Borntraeger, Berlin, Stuttgart (Lehrbuch der Hydrogeologie, 9).

Kendall, C.; McDonnell, J. J. (ed.) (1998): Isotope Tracers in Catchment Hydrology. Elsevier, Amsterdam.

Kenner, R.; Chinellato, G.; Iasio, C.; Mosna, D.; Cuozzo, G.; Benedetti, E.; Visconti, M. G.; Manunta, M.; Phillips, M.; Mair, V.; Zischg, A.; Thiebes, B.; Strada, C. (2016): Integration of space-borne DInSAR data in a multi-method monitoring concept for alpine mass movements. Cold Regions Science and Technology 131, 65–75.

Kenner, R.; Pruessner, L.; Beutel, J.; Limpach, P.; Phillips, M. (2020): How rock glacier hydrology, deformation velocities and ground temperatures interact. Examples from the Swiss Alps. Permafrost Periglac. Process. 31 (1), 3–14.

King, L.; Fisch, W.; Haeberli, W.; Waechter, H. P. (1987): Comparison of resistivity and radio-echo soundings on rock glacier permafrost. Zeitschrift für Gletscherkunde und Glazialgeologie 23 (1), 77–97.

King, L. (1990): Soil and Rock Temperatures in Discontinuous Permafrost: Gornergrat and Unterrothorn, Wallis, Swiss Alps. Permafrost Periglac. Process. 1, 177–188.

Klug, C.; Bollmann, E.; Rieg, L.; Sproß, M.; Sailer, R.; Stötter, J. (2014): Detecting and Quantifying Area Wide Permafrost Change. In: Rutzinger, M.; Heinrich, K.; Borsdorf, A.; Stötter, J. (ed.): permAfrost—Austrian Permafrost Research Initiative. Final Report. Österreichische Akademie der Wissenschaften (ÖAW), Wien (IGF Forschungsberichte, 6), 68–108.

Klug, C. (2015): Blockgletscherbewegung im Äußeren Hochebenkar 1953–2010 – eine Methodenkombination aus digitaler Photogrammetrie und Airborne Laserscanning. In: Schallhart, N.; Erschbamer, B. (ed.): Forschung am Blockgletscher. Methoden und Ergebnisse. Innsbruck Univ. Press, Innsbruck (Alpine Forschungsstelle Obergurgl, 4), 135–158.

Konrad, S. K.; Humphrey, N. F.; Steig, E. J.; Clark, D. H.; Potter, N.; Pfeffer, W. T. (1999): Rock glacier dynamics and paleoclimatic implications. Geology 27 (12), 1131.

Konrad, S. K.; Humphrey, N. F. (2000): Steady-state flow model of debris-covered glaciers (rock glaciers). In: Fountain, A.; Raymond, C. F.; Nakao, M. (ed.): Debris-covered Glaciers. Proceedings of an International Workshop Held at the University of Washington in Seattle, Washington, USA, 13–15 September 2000. International Association of Hydrological Sciences (IAHS) Press, 255–263.

Kovács, A.; Perrochet, P.; Király, L.; Jeannin, P.-Y. (2005): A quantitative method for the characterisation of karst aquifers based on spring hydrograph analysis. Journal of Hydrology 303 (1-4), 152–164.

Krainer, K.; Mostler, W. (2000): Reichenkar rock glacier. A glacier derived debris-ice system in the western Stubai Alps, Austria. Permafrost Periglac. Process. 11 (3), 267–275.

Krainer, K.; Mostler, W. (2002): Hydrology of Active Rock Glaciers. Examples from the Austrian Alps. Arctic, Antarctic, and Alpine Research 34 (2), 142–149.

Krainer, K.; Mostler, W. (2006): Flow Velocities of Active Rock Glaciers in the Austrian Alps. Geografiska Annaler. Series A, Physical Geography 88 (4), 267–280.

Krainer, K.; Mostler, W.; Spötl, C. (2007): Discharge from active rock glaciers, Austrian Alps: A stable isotope approach. AJES 100, 102–112.

Krainer, K.; Lang, K.; Hausmann, H. (2010): Active Rock Glaciers at Croda Rossa/Hohe Gaisl, Eastern Dolomites (Alto Adige/South Tyrol, Northern Italy). Geogr. Fis. Dinam. Quat. (33), 25–36.

Krainer, K.; Mussner, L.; Behm, M.; Hausmann, H. (2012): Multi-disciplinary investigation of an active rock glacier in the Sella Group (Dolomites; Northern Italy). AJES 105 (2), 48–62.

Krainer, K. (2015): Der aktive Blockgletscher im Äußeren Hochebenkar. In: Schallhart, N.; Erschbamer, B. (ed.): Forschung am Blockgletscher. Methoden und Ergebnisse. Innsbruck: Innsbruck Univ. Press (Alpine Forschungsstelle Obergurgl, 4), 55–75.

Krainer, K.; Bressan, D.; Dietre, B.; Haas, J. N.; Hajdas, I.; Lang, K.; Mair, V.; Nickus, U.; Reidl, D.; Thies, H.; Tonidandel, D. (2015): A 10,300-year-old permafrost core from the active rock glacier Lazaun, southern Ötztal Alps (South Tyrol, northern Italy). Quaternary Research 83 (2), 324–335.

Kreft, A.; Zuber, A. (1978): On the physical meaning of the dispersion equation and its solutions for different initial and boundary conditions. Chemical Engineering Science 33 (11), 1471–1480.

Kresic, N.; Stevanovic, Z. (Hg.) (2010): Groundwater Hydrology of Springs. Engineering, Theory, Management, and Sustainability, Burlington, MA. Elsevier, Butterworth-Heinemann.

Kurylyk, B. L.; Watanabe, K. (2013): The mathematical representation of freezing and thawing processes in variably-saturated, non-deformable soils. Advances in Water Resources 60, 160–177.

Langevin, C. D.; Hughes, J. D.; Banta, E. R.; Niswonger, R. G.; Panday, S.; Provost, A. (2017): Documentation for the MODFLOW 6 Groundwater Flow Model. U. S. Geological Survey (USGS), Reston, Va. (U. S. Geol. Surv. Techniques and Methods, 6-A55).

Leibundgut, C.; Maloszewski, P.; Külls, C. (2009): Tracers in hydrology. Wiley-Blackwell, Chichester.

Leopold, M.; Williams, M. W.; Caine, N.; Völkel, J.; Dethier, D. (2011): Internal structure of the Green Lake 5 rock glacier, Colorado Front Range, USA. Permafrost Periglac. Process. 22 (2), 107–119.

Liaudat, D. T.; Sileo, N.; Dapeña, C. (2020): Periglacial water paths within a rock glacier-dominated catchment in the Stepanek area, Central Andes, Mendoza, Argentina. Permafrost Periglac. Process. 26 (1), 175.

Liedl, R.; Sauter, D.; Hückinghaus, M.; Clemens, T.; Teutsch, G. (2003): Simulation of the development of karst aquifers using a coupled continuum pipe flow model. Water Resour. Res. 39 (1).

Liu, F.; Williams, M. W.; Caine, N. (2004): Source waters and flow paths in an alpine catchment, Colorado Front Range, United States. Water Resour. Res. 40 (9), 61.

Lomb, N. R. (1976): Least-squares frequency analysis of unequally spaced data. Astrophys Space Sci 39 (2), 447–462.

Luetschg, M.; Stoeckli, V.; Lehning, M.; Haeberli, W.; Ammann, W. (2004): Temperatures in two boreholes at Flüela Pass, Eastern Swiss Alps. The effect of snow redistribution on permafrost distribution patterns in high mountain areas. Permafrost Periglac. Process. 15 (3), 283–297.

Luetschg, M.; Lehning, M.; Haeberli, W. (2008): A sensitivity study of factors influencing warm/thin permafrost in the Swiss Alps. J. Glaciol. 54 (187), 696–704.

Maillet, E. (1905): Essai d'hydraulique souterraine et fluviale. Librairie scientifique A. Hermann, Paris.

Maraqa, M. A. (2001): Prediction of mass-transfer coefficient for solute transport in porous media. Journal of Contaminant Hydrology 50 (1-2), 1–19.

Mari, S.; Scapozza, C.; Pera Ibarguren, S.; Delaloye, R. (2013): Prove di multitracciamento di ghiacciai rocciosi e ambienti periglaciali nel Vallon de Réchy (VS) e nella Valle di Sceru (TI). Bollettino della Società ticinese di Scienze naturali 101, 13–20.

Mayaud, C.; Walker, P.; Hergarten, S.; Birk, S. (2015): Nonlinear Flow Process. A New Package to Compute Nonlinear Flow in MODFLOW. Groundwater 53 (4), 645–650.

Mewes, B.; Hilbich, C.; Delaloye, R.; Hauck, C. (2017): Resolution capacity of geophysical monitoring regarding permafrost degradation induced by hydrological processes. The Cryosphere 11 (6), 2957–2974.

Millington, R. J.; Quirk, J. P. (1961): Permeability of porous solids. Trans. Faraday Soc. 57, 1200–1206.

Mohammed, A. A.; Kurylyk, B. L.; Cey, E. E.; Hayashi, M. (2018): Snowmelt Infiltration and Macropore Flow in Frozen Soils. Overview, Knowledge Gaps, and a Conceptual Framework. Vadose Zone Journal 17 (1).

Monnier, S.; Kinnard, C. (2015): Internal Structure and Composition of a Rock Glacier in the Dry Andes, Inferred from Ground-penetrating Radar Data and its Artefacts. Permafrost Periglac. Process. 26 (4), 335–346.

Mull, D. S.; Liebermann, T. D.; Smoot, J. L.; Woosley, L. H., Jr. (1988): Application of Dye-Tracing Techniques for Determining Solute-Transport Characteristics of Ground Water in Karst Terranes. EPA904/6–88–001. United States Environmental Protection Agency (EPA) und United States Geological Survey (USGS), Atlanta, GA.

Musil, M.; Maurer, H.; Hollinger, K.; Green, A. G. (2006): Internal structure of an alpine rock glacier based on crosshole georadar traveltimes and amplitudes. Geophys. Prospect. 54 (3), 273–285.

Narasimhan, T. N.; Witherspoon, P. A. (1976): An Integrated Finite Difference Method for Analyzing Fluid Flow in Porous Media. Water Resour. Res. 12 (1), 57–64.

Nye, J. F. (1952): The Mechanics of Glacier Flow. Journal of Glaciology 2 (12), 82–93.

Nye, J. F. (1953): The flow law of ice from measurements in glacier tunnels, laboratory experiments and the Jungfraufirn borehole experiment. Proc. R. Soc. Lond. A 219, 477–489.

Nye, J. F. (1957): The distribution of stress and velocity in glaciers and ice-sheets. Proc. R. Soc. Lond. A 239 (1216), 113–133.

Panday, S.; Langevin, C. D.; Niswonger, R. G.; Ibaraki, M.; Hughes, J. D. (2013): MOD-FLOW–USG Version 1: An Unstructured Grid Version of MODFLOW for Simulating Groundwater Flow and Tightly Coupled Processes Using a Control Volume Finite-Difference Formulation. U. S. Geological Survey (USGS), Reston, Va. (U. S. Geol. Surv. Techniques and Methods, 6-A45).

Pang, L.; Close, M. (1999): Field-scale physical non-equilibrium transport in an alluvial gravel aquifer. Journal of Contaminant Hydrology 38 (4), 447–464.

Parker, J. C.; van Genuchten, M. T. (1984): Flux-Averaged and Volume-Averaged Concentrations in Continuum Approaches to Solute Transport. Water Resour. Res. 20 (7), 866–872.

Pauritsch, M. (2011): Die Hydrodynamik reliktischer Blockgletscher am Beispiel des Schönebenblockgletschers (Seckauer Tauern, Steiermark). Master Thesis. Karl-Franzens-Universität, Graz.

Pauritsch, M.; Birk, S.; Wagner, T.; Hergarten, S.; Winkler, G. (2015): Analytical approximations of discharge recessions for steeply sloping aquifers in alpine catchments. Water Resour. Res. 51 (11), 8729–8740.

Pauritsch, M.; Wagner, T.; Winkler, G.; Birk, S. (2017): Investigating groundwater flow components in an Alpine relict rock glacier (Austria) using a numerical model. Hydrogeology Journal 25 (2), 371–383.

Pedevilla, T. (2019): Geologische, geomorphologische und hydrogeologische Untersuchungen an Blockgletschern im Lareintal (Tirol, Österreich). Master Thesis, University of Innsbruck.

Perfect, E.; Williams, P. J. (1980): Thermally induced water migration in frozen soils. Cold Regions Science and Technology 3, 101–109.

Potter, N., Jr. (1969): Rock glaciers and mass-wastage in the Galena Creek area, northern Absaroka Mountains, Wyoming, Doctoral Thesis, University of Minnesota.

Potter, N. (1972): Ice-Cored Rock Glacier, Galena Creek, Northern Absaroka Mountains, Wyoming. Geological Society of America Bulletin 83 (10), 3025.

Provost, A.; Langevin, C. D.; Hughes, J. D. (2017): Documentation for the "XT3D" Option in the Node Property Flow (NPF) Package of MODFLOW 6. U. S. Geological Survey (USGS), Reston, Va. (U. S. Geol. Surv. Techniques and Methods, 6-A56).

R Development Core Team (2019): R. A language and environment for statistical computing. Version 36, Wien. R Foundation for Statistical Computing. http://www.R-project.org.

Rajczak, J.; Pall, P.; Schär, C. (2013): Projections of extreme precipitation events in regional climate simulations for Europe and the Alpine Region. J. Geophys. Res. Atmos. 118 (9), 3610–3626.

Rehrl, C.; Birk, S. (2010): Hydrogeological Characterisation and Modelling of Spring Catchments in a Changing Environment. AJES 103 (2), 106–117.

Reimann, T.; Rehrl, C.; Shoemaker, W. B.; Geyer, T.; Birk, S. (2011): The significance of turbulent flow representation in single-continuum models. Water Resour. Res. 47 (9).

Rieder, A. (2017): Geologische, geomorphologische und hydrogeologische Untersuchungen im Bereich Ölgrube, Kaunergrat, Ötztaler Alpen. Master Thesis. Universität Innsbruck, Innsbruck.

Roer, I.; Haeberli, W.; Avian, M.; Kaufmann, V.; Delaloye, R.; Lambiel, C.; Kääb, A. (2008): Observations and Considerations on Destabilizing Active Rock Glaciers in the European Alps. In: Kane, D. L.; Hinkel, K. M. (ed.): Proceedings of the 9th International Conference on Permafrost. University of Alaska Fairbanks, 1505–1510.

Rogger, M.; Chirico, G. B.; Hausmann, H.; Krainer, K.; Brückl, E.; Stadler, P.; Blöschl, G. (2017): Impact of mountain permafrost on flow path and runoff response in a high alpine catchment. Water Resour. Res. 53.

Ruf, T. (2019): Lomb. R package version 1.2.

Sahuquillo, A. (1983): An Eigenvalue Numerical Technique for Solving Unsteady Linear Groundwater Models Continuously in Time. Water Resour. Res. 19 (1), 87–93.

Scargle, J. D. (1982): Studies in astronomical time series analysis. II—Statistical aspects of spectral analysis of unevenly spaced data. The Astrophysical Journal 263, 835-853.

Schnegg, P.-A. (2002): An inexpensive field fluorometer for hydrogeological tracer tests with three tracers and turbidity measurement. In: Bocanegra, E.; Martínez, D.; Massone, H. (ed.): XXXII IAH and ALHSUD Congress Groundwater and Human Development, Mar del Plata, Argentina, October 2002, 1484–1488.

Schulze-Makuch, D.; Carlson, D. A.; Cherkauer, D. S.; Malik, P. (1999): Scale Dependency of Hydraulic Conductivity in Heterogeneous Media. Ground Water 37 (6), 904–919.

Seelig, M. (in prep.): Analysis of tracer tests at a rock glacier (Schöneben, Austria) using a numerical transport model. Master Thesis, University of Graz.

Shoemaker, W. B.; Kuniansky, E. L.; Birk, S.; Bauer, S.; Swain, E. D. (2008): Documentation of a Conduit Flow Process (CFP) for MODFLOW-2005. U. S. Geological Survey (USGS), Reston, Virginia (Techniques and Methods, Book 6, Chapter A24).

Smart, P. L.; Laidlaw, I. M. S. (1977): An Evaluation of Some Fluorescent Dyes for Water Tracing. Water Resour. Res. 13 (1), 15–33.

Springman, S. M.; Arenson, L. U.; Yamamoto, Y.; Maurer, H.; Kos, A.; Buchli, T.; Derungs, G. (2012): Multidisciplinary investigations on three rock glaciers in the swiss alps. Legacies and future perspectives. Geografiska Annaler: Series A, Physical Geography 94 (2), 215–243.

Stähli, M.; Jansson, P.-E.; Lundin, L.-C. (1996): Preferential Water Flow in a Frozen Soil—A Two-Domain Model Approach. Hydrol. Process. 10 (10), 1305–1316.

Stähli, M.; Jansson, P.-E.; Lundin, L.-C. (1999): Soil moisture redistribution and infiltration in frozen sandy soils. Water Resour. Res. 35 (1), 95–103.

Taylor, G. I. (1954): The dispersion of matter in turbulent flow through a pipe. Proc. R. Soc. Lond. A 223 (1155), 446–468.

Tenthorey, G. (1992): Perennial névés and the hydrology of rock glaciers. Permafrost Periglac. Process. 3 (3), 247–252.

Tenthorey, G. (1993): Paysage géomorphologique du Haut-Val de Réchy (Valais, Suisse) et hydrologie liée aux glaciers rocheux. Dissertation. Université de Fribourg, Fribourg. Institut de Géographie.

Tenthorey, G. (1994): Hydrologie liee aux glaciers rocheux, Haut-Val de Réchy (Nax, VS). Bull. Murithienne 112, 97–116.

Thies, H.; Nickus, U.; Tolotti, M.; Tessadri, R.; Krainer, K. (2013): Evidence of rock glacier melt impacts on water chemistry and diatoms in high mountain streams. Cold Regions Science and Technology 96, 77–85.

Toride, N.; Leij, F. J.; van Genuchten, M. T. (1993): A Comprehensive Set of Analytical Solutions for Nonequilibrium Solute Transport With First-Order Decay and Zero-Order Production. Water Resour. Res. 29 (7), 2167–2182.

Toride N.; Leij, F. J.; van Genuchten, M. T. (1999): The CXTFIT Code for Estimating Transport Parameters from Laboratory or Field. Tracer Experiments. Version 2.1. U.S. Salinity Laboratory Agricultural Research Service.

US National Aeronautics and Space Administration (NASA); US National Geospatial-Intelligence Agency (NGA) (2009): Shuttle Radar Topography Mission 3 Arc-Second Data (SRTM-3), v2.1. https://dds.cr.usgs.gov/srtm/version2_1, last accessed 20.11.2020.

Untersweg, T.; Proske, H. (1996): Untersuchungen an einem fossilen Blockgletscher im Hochreichhartgebiet (Niedere Tauern, Steiermark). Grazer Schriften Geogr Raumfor 33, 201–207.

Untersweg, T.; Schwendt, A. (1995): Die Quellen der Blockgletscher in den Niederen Tauern. Amt der Steiermärkischen Landesregierung, Graz (Berichte der wasserwirtschaftlichen Rahmenplanung, 78).

Untersweg, T.; Schwendt, A. (1996): Blockgletscher und Quellen in den Niedern Tauern. Mitt. Österr. Geol. Ges. 87, 47–55.

van Genuchten, M. T.; Wagenet, R. J. (1989): Two-Site/Two-Region Models for Pesticide Transport and Degradation: Theoretical Development and Analytical Solutions. Soil Sci. Soc. Am. J. 53 (5), 1303–1310.

Vick, S. G. (1981): Morphology and the role of landsliding in formation of some rock glaciers in the Mosquito Range, Colorado. Geological Society of America Bulletin 92 (1), 75–84.

Vonder Mühll, D. S. (1993): Geophysikalische Untersuchungen im Permafrost des Oberengadins. Doctoral Thesis. Eidgenössische Technische Hochschule, Zürich.

Vonder Mühll, D. S.; Arenson, L. U.; Springman, S. M. (2003): Temperature conditions in two Alpine rock glaciers. In: Phillips, M.; Springman, S. M.; Arenson, L. U. (ed.): Proceedings of the 8th International Conference on Permafrost. Vol. 2. Balkema, Lisse, 1195–1200.

Wagner, T.; Mayaud, C.; Benischke, R.; Birk, S. (2013): Ein besseres Verständnis des Lurbach-Karstsystems durch ein konzeptionelles Niederschlags-Abfluss-Modell. Grundwasser 18 (4), 225–235.

Wagner, T.; Pauritsch, M.; Winkler, G. (2016): Impact of relict rock glaciers on spring and stream flow of alpine watersheds. Examples of the Niedere Tauern Range, Eastern Alps (Austria). AJES 109 (1).

Wagner, T.; Seelig, S.; Wedenig, M.; Pleschberger, R.; Krainer, K.; Kellerer-Pirklbauer, A.; Ribis, M.; Hergarten, S.; Winkler, G. (2019a): Wasserwirtschaftliche Aspekte von Blockgletschern in Kristallingebieten der Ostalpen. Speicherverhalten, Abflussdynamik und Hydrochemie mit Schwerpunkt Schwermetallbelastungen. Bundesministerium für Nachhaltigkeit und Tourismus (BMNT), Wien.

Wagner, T.; Pauritsch, M.; Mayaud, C.; Kellerer-Pirklbauer, A.; Thalheim, F.; Winkler, G. (2019b): Controlling factors of microclimate in blocky surface layers of two nearby relict rock glaciers (Niedere Tauern Range, Austria). Geografiska Annaler: Series A, Physical Geography 105 (2), 1–24.

Wagner, T.; Pleschberger, R.; Seelig, S.; Ribis, M.; Kellerer-Pirklbauer, A.; Krainer, K.; Philippitsch, R.; Winkler, G. (2020a): The first consistent inventory of rock glaciers and their hydrological catchments of the Austrian Alps. AJES 113 (1), 1–23.

Wagner, T.; Brodacz, A.; Krainer, K.; Winkler, G. (2020b): Active rock glaciers as shallow groundwater reservoirs, Austrian Alps. Grundwasser—Zeitschrift der Fachsektion Hydrogeologie 37 (1), 13.

Wagner, T.; Ribis, M.; Kellerer-Pirklbauer, A.; Krainer, K.; Winkler, G. (2020c): The Austrian rock glacier inventory RGI_1 and the related rock glacier catchment inventory RGCI_1 in ArcGis (shapefile) format. PANGAEA, https://doi.org/10.1594/PANGAEA. 921629, last accessed 20.11.2020.

Wagner, T.; Seelig, S.; Krainer, K.; Winkler, G. (in prep.): Storage-discharge characteristics of an active rock glacier catchment in the Innere Ölgrube, Austrian Alps.

Wahrhaftig, C.; Cox, A. (1959): Rock Glaciers in the Alaska Range. Geological Society of America Bulletin 70 (4), 383.

Watkins, J. S.; Walters, L. A.; Godson, R. H. (1972): Dependence of in-situ compressional wave velocity on porosity in unsaturated rocks. Geophysics 37 (1), 29–35.

Whalley, W. B.; Martin, H. E. (1992): Rock glaciers. II models and mechanisms. Progress in Physical Geography 16 (2), 127–186.

White, S. E. (1971): Rock Glacier Studies in the Colorado Front Range, 1961 to 1968. Arctic and Alpine Research 3 (1), 43–64.

Williams, M. W.; Knauf, M.; Caine, N.; Liu, F.; Verplanck, P. L. (2006): Geochemistry and source waters of rock glacier outflow, Colorado Front Range. Permafrost Periglac. Process. 17 (1), 13–33.

Winkler, G.; Wagner, T.; Pauritsch, M.; Birk, S.; Kellerer-Pirklbauer, A.; Benischke, R.; Leis, A.; Morawetz, R.; Schreilechner, M. G.; Hergarten, S. (2016a): Identification and assessment of groundwater flow and storage components of the relict Schöneben Rock Glacier, Niedere Tauern Range, Eastern Alps (Austria). Hydrogeol J 24 (4), 937–953.

Winkler, G.; Pauritsch, M.; Wagner, T.; Kellerer-Pirklbauer, A. (2016b): Reliktische Blockgletscher als Grundwasserspeicher in alpinen Einzugsgebieten der Niederen Tauern. Das Land Steiermark, Graz (Berichte der Wasserwirtschaftlichen Planung Steiermark, 87).

Winkler, G.; Wagner, T.; Krainer, K.; Ribis, M.; Hergarten, S. (2018a): Hydrogeology of Rock Glaciers—Storage Capacity and Drainage Dynamics—An Overview. In: Sychev, V. G.; Mueller, L. (ed.): Novel methods and results of landscape research in Europe, Central Asia and Siberia. Vol II/71. Moskau: Russian Academy of Sciences, FSBSI All-Russian Research Institute of Agrochemistry, 329–334.

Winkler, G.; Seelig, S.; Wagner, T.; Leis, A.; Krainer, K. (2018b): Flow component characterization of alpine aquifers—an isotopic approach. 16th Stable Isotope Network Meeting (SINA), Graz.

Wirz, V.; Gruber, S.; Purves, R. S.; Beutel, J.; Gärtner-Roer, I.; Gubler, S.; Vieli, A. (2016): Short-term velocity variations at three rock glaciers and their relationship with meteorological conditions. Earth Surf. Dynam. 4 (1), 103–123.

Zenklusen Mutter, E.; Phillips, M. (2012): Active Layer Characteristics At Ten Borehole Sites In Alpine Permafrost Terrain, Switzerland. Permafrost Periglac. Process. 23 (2), 138–151.

Zhang, T. (2005): Influence of the seasonal snow cover on the ground thermal regime. An overview. Rev. Geophys. 43 (4), 1.

Zhou, X.; Buchli, T.; Kinzelbach, W.; Stauffer, F.; Springman, S. M. (2015): Analysis of Thermal Behaviour in the Active Layer of Degrading Mountain Permafrost. Permafrost Periglac. Process. 26 (1), 39–56.

Żurawek, R. (2002): Internal Structure of a relict rock glacier, Sleza Massif, Southwest Poland. Permafrost Periglac. Process. 13 (1), 29–42.

Żurawek, R. (2003): The problem of identification of relict rock glaciers on sedimentological evidence. Landform Analysis 4, 7–15.

Printed in the United States
by Baker & Taylor Publisher Services